The Enigma of Sunspots

Judit Brody

The Enigma of Sunspots

A story of discovery and
scientific revolution

Floris Books

First published in 2002 by Floris Books

British Library CIP Data available

ISBN 0-86315-370-4

Printed in Poland

Contents

Acknowledgments

My thanks to Gail Vines who many years ago encouraged me to start writing, and to Charles O'Reilly who read the first few pages of this book and encouraged me to carry on. I am grateful to Geoff Elston of the British Astronomical Association for advice on how to observe the sun. Thanks also to my son Michael Kovari and my editor Christopher Moore who helped me put the book into shape, and to Hannah Perry for help with the illustrations. The staff of the Science Museum Library and the British Library provided me with most of the material and I am deeply indebted to them for their assistance.

∾ 1 ⊌

Gazers and Gawpers

We need its light, we need its heat, we need its energy,
Without the Sun without a doubt, there'd be no you and me.

This childish couplet appropriately captures the plain fact, known from the dawn of history, that the Sun sustains all life on Earth. Naturally enough, for thousands of years many cultures venerated the Sun for giving us light and heat. Rituals and festivities made sure that it did not fail to rise in the morning and would not disappear suddenly from the sky. In the middle of winter when the days begin to lengthen, special celebrations were (and still are) called for.

The challenge of discovering the secrets of the Sun has kept scientists busy during the last three thousand years. They carefully watched the progression of the Sun in the sky, measured its distance from the Earth, analysed its rays and tried to detect its features. They put forward theories and explanations of how it produces an enormous quantity of heat and yet does not seem to burn out. During that time we have learned much, but in some ways the real essence of the Sun, the nearest star to Earth, is still a matter of conjecture. 'The Sun is stranger than you think, displaying mysterious manifestations of the familiar laws of physics and posing new problems with every major advance' was the recent verdict of Eugene Parker, one of the foremost American astrophysicists.[1]

At the beginning of the seventeenth century four astronomers claimed to have discovered dark blotches appearing occasionally on the surface of the Sun: Johann Fabricius, Christoph Scheiner, Galileo Galilei and Thomas Harriot. This discovery and the accurate recording of the shapes and positions of the spots gave astronomers their very first opportunity to find out something about the nature of the Sun. The spots provided evidence for the first time that the Sun is a spherical body and not just a flat disc in the sky. Four hundred years ago when regular observations began, sunspots also helped demolish ancient ideas of a perfect and unchangeable heaven made from an entirely different substance from the Earth and anything on it, and obeying entirely different physical laws. Of course, this did not happen without controversy and the rational deductions and conjectures on what causes the spots, what they really are and what their influence here on Earth might be, have been matched by some of the wildest speculations human imagination can conjure up.

The brightest object in the sky always fascinated the curious and in the more distant past the Sun and its spots were occasionally observed with the naked eye. How was this done without those wonderful protective sunglasses we can now buy to observe an eclipse? Reflection provides a simple method to view the Sun's disc without harming one's eyes. In calm weather the surface of a pond or a lake is like a mirror. In mythology Narcissus was tempted to admire his own reflection in such a mirror. It may have produced a better picture than mirrors available at the time. Similarly, the rays of the Sun are partially reflected on the surface of water and it is more than likely that the Chinese used this observational method. I watched part of the solar eclipse of 1999 sitting in an outdoor café with my family and looking at the mirror image of the waning Sun as it appeared on the highly polished table in front of us.

But some of the ancient observers thought that they knew how to attenuate the rays and gazed through partially opaque objects. The Chinese used thin slices of jade and other semi-transparent minerals for this purpose. Then there were others who squinted at the Sun with the naked eye when it was not too uncomfortable to do so. In hazy or lightly cloudy conditions it is quite possible to look into the Sun for a short period. And in the early morning and late evening when the Sun is low above the horizon, it shines through a thick layer of air which attenuates the rays.* Did these observers know that their eyes could suffer permanent damage? In the early days they probably did not. Neither the nature of light and heat, nor the vulnerability of the human eye was properly understood. Pain and discomfort were not always associated with lasting physical damage either and the invention of the telescope brought dangers that were at first unrecognized. Some of the early observers projected the disc of the Sun onto a piece of paper not because they wished to preserve their eyesight but rather because they found looking straight through the instrument much too painful.

It had been known that sailors who had to look at the Sun when it was high in the sky and in all weathers, often had poor eyesight and occasionally went blind but their ailments could have been caused by any number of reasons, such as for instance salty water getting into their eyes, or by poor nutrition on board ship.† Sun spotters (as they are sometimes called) watched in hazy conditions when the Sun was rising or setting. In my childhood we were busy preparing smoked pieces of glass when an eclipse was expected. Looking at the Sun through smoked glass, now held to be a dangerous practice, was perfectly acceptable sixty years ago.

Ignorance of possible harm and ignorance of how likely that harm might be are only two of the many reasons why people take risks. Galileo did know from personal experience that peering at the Sun through a telescope for long periods will result in 'fatigue and injury to the eyes.' Yet he did not entirely avoid looking at the Sun and his blindness in old age has been attributed in part to keen sky watching. Would he have stopped had he known what was going to happen?

In 1691 in a letter to John Locke the philosopher, Sir Isaac Newton recalled an experiment he once made

> with ye hazzard of my eyes. The manner was this. I looked a very little while upon ye Sun in a looking–glass with my right eye & then turned my eyes into a dark corner of my chamber & winked to observe the impression made & the

Sir Isaac Newton

* It cannot be stressed strongly enough that none of these methods are recommended to those who wish to keep their eyesight.

† The backstaff, invented in 1595, enabled its user to measure the position of the Sun while looking away from it.

circles of colours which encompassed it & how they de-
cayed by degrees & at last vanished. This I repeated a second & a third time ... in a few hours time I had brought my
eys to such a pass that I could look upon no bright object
with either eye but I saw the Sun before me, so that I durst
neither write nor read but to recover ye use of my eyes
shut my self up in my chamber made dark for three days.[2]

Although he is talking about 'hazzard' to his eyes, Newton
attributed the sensations more to psychological than to physio-
logical effects. Would he have experimented had he known that
he could permanently harm his eyes? We do not know. After all,
Newton is known for poking bodkins into his own eyes during
another experiment and he must have realized how dangerous
that was. Similarly, physicians in the past often experimented by
taking drugs or performing surgical procedures on themselves
although they were well aware of the possible harm they might
suffer. Curiosity, the wish to find out and to know — occasionally driven by a desire to benefit humankind — the thrill and
excitement of a chase, the promise of honour, glory, and power
and the hope of lasting fame, these are some reasons why people
are willing to take risks knowingly.

Attitudes to risk and to perceived risk vary from culture to
culture. We now live relatively sheltered lives and are increasingly
reluctant to take risks, although our perception of how risky an
action might be is often fallacious. There are other factors in
operation as well. In 1999, possibly driven partly by fear of liti-
gation, officialdom issued strict warnings in advance of the
eclipse. We were told not to look into the Sun at all, not even during the total phase of the eclipse. We were not to use the special
spectacles in case they had a tiny hole in them. Astronomers
rebelled. They thoroughly checked their eclipse viewers and were
shown on television looking into the full Sun with their shades
on. Were they taking a calculated risk?

Our story, or rather history, of sunspots is not unlike a detec-
tive mystery: small crumbs of information and occasional unexpected insights finally add up to a solution. There is a difference

THE ENIGMA OF SUNSPOTS

though: no single investigator, no Monsieur Poirot or Miss Marple comes up with the answer. The threads are multifarious and the motley crew of detectives numerous. Over the years many, who had some kind of training in astronomy and whose livelihood in some way depended on it, turned their eyes to the Sun. Watching the nearest star were also men of independent means, who devoted their entire lives, and their money, to observing the sky: the 'grand amateurs.' So were many others for whom astronomy was only a leisure activity. Both the grand amateurs and the spare time sky watchers have made significant contributions to our knowledge. Chemistry and physics do not, and never did, muster such an army of amateurs as astronomy, although in the nineteenth century geology and botany were also popular hobbies.

Astronomy attracted women from the earliest times. They appear at first as helpmates to their husbands or brothers, but later as amateur and professional astronomers in their own right. We are going to meet some of these women to whom other branches of the sciences were often closed but astronomical work gave an opportunity. By the nineteenth century the amateur astronomical societies numbered many women among their members and expositions of astronomy written by women authors became popular. Here are a few examples. Written for children, *A Comprehensive Astronomical and Geographical Classbook* (1815) and *An Astronomical Catechism* (1818) by the teachers Mrs Margaret Bryan and by Catherine Vale Whitwell respectively, were popular teaching aids and the latter includes a discussion of sunspots. (Mrs Whitwell drew her own illustrations of sunspots.) Mrs Winifred Lockyer, the first wife of the astronomer Sir Joseph Norman Lockyer was a leading translator of French treatises on astronomy and

Margaret Bryan and children. From A Compendious System of Astronomy *(1797)*

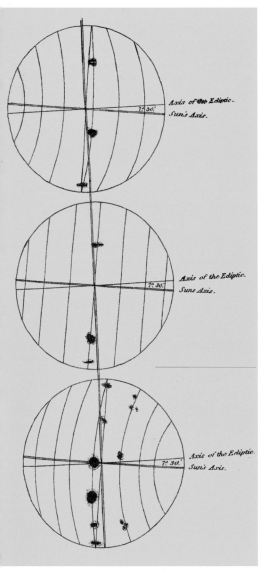

Catherine Vale Whitwell's drawings from her Astronomical Catechism.

on physics. Mary Somerville's translation from the French of Pierre Laplace's *Mechanism of the Heavens* (1831), and her own *On the Connexion of the Physical Sciences* (1834) were both highly regarded. Neither of these women had any formal training in astronomy. They attended the occasional free lecture and picked up whatever they could from family, friends, and books on the subject.

One of the best historians of astronomy of all time was Agnes Clerke. She was lucky in that her bank manager father was really a scientist at heart and that his enthusiasm extended to teaching his two daughters all he knew about scientific matters. They watched the sky together, and at weekends the whole neighbourhood smelled from the chemical experiments they performed in the kitchen. Both daughters became respected authors but Agnes Clerke was an especially serious and prolific author on a variety of topics. That she was commissioned to contribute to several reference works is proof that her talents were recognized. Her book on the history of astronomy during the nineteenth century went into several editions and is still one of the best source books of the period. For her work Clerke received prizes and honours from scientific societies (like Mary Somerville before her).

But anyone who thinks that it was the romantic setting of gazing at the sky under cloak of darkness that enticed women to the · study of astronomy, should think again: many of them were diligent

THE ENIGMA OF SUNSPOTS

sunspot watchers. Elizabeth Brown, one of the founder members of the British Astronomical Association and first head of its solar section, recommended sunspot watching and drawing as more appropriate for women than studying the stars because no exposure to the night air was involved. Keeping in mind the frequency of tuberculosis among young women, this is not such a strange Victorian notion as one might think at first.

As soon as telescopes became more affordable, amateur activity really took off. The relative freedom to take up new ideas and to make observations and experiments without being accountable to a boss or to an institution led to great discoveries by amateurs. Even today a remark of Mrs Huggins, the wife of a Victorian amateur astronomer and an amateur astronomer herself, is still valid: 'Astronomical professorships and official observatories may multiply, but ... the day of the amateur astronomer is not past.'[3] Spacecraft observations, Hubble telescopes and space probes notwithstanding, amateur astronomers are still active. Amateur sunspot watchers contribute to a data bank, amateur astronomical clubs and societies flourish and audience figures for popular astronomical programmes remain consistently high. A glance at the display on the newsagents' shelves will also reveal the unceasing interest the general public has in the sky — though admittedly somewhat less than in body-building or in slimming diets.

Above: *Agnes Clerke*
Below: *Drawings by Elizabeth Brown, first director of the Solar Section of the British Astronomical Association.*

Although in these days high above the atmosphere, spacecraft and artificial satellites carry sophisticated instruments probing every imaginable physical and chemical property of the Sun, straightforward imaging of the Sun and its spots remains important. Solar scientists and astronomers, using methods that do not harm the eye, record the number of sunspots and sunspot groups every day and study the changes they visibly undergo in the

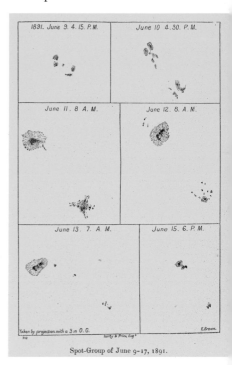

Spot-Group of June 9-17, 1891.

Solar cycles 21–23.
The thick line is the
monthly smoothed
sunspot number
(SSN)while the thin
line is the actual
monthly sunspot
number.
Cycle 21 started in
June 1976 with an
SSN of 12.2.
Cycle 22 started in
September 1986
with SSN 12.3.
Cycle 23 started in
May 1996 with the
monthly SSN at 8.0
and peaked in April
2000 at 120.8.
The last SSN is for
September 2001
and is 114.1.
The next solar
minimum will
probably occur
sometime in 2006
with cycle 24
peaking in 2010.

Data source:
Sunspot Index Data
Center in Brussels.
www.dxlc.com/
solar/solcycle.html

Solar Cycle 21

hope that the spots' behaviour will eventually reveal more of the deeper secrets of the Sun. Because 'these troublesome spots' (as Galileo called them) are the results of physical processes taking place in the Sun's interior, they continue to hold one of the most important keys in the pursuit of a better understanding of our nearest star.

There are still many unanswered questions. Why is the face of the Sun more freckled at certain times and less so at other times? About every eleven years the number of sunspots reaches a maximum, then it starts to fall, eventually to rise again to a new maximum. 2000 was a year of such a maximum and for solar scientists this was more important than the fact that it was declared the millennial year. What causes this cycle, why is it not entirely regular, and why are there other, longer periods as well when spots disappear for several years? Why do most spots form only on certain fairly well defined parts of the solar disc? What is the process that makes some spots live longer than others? What is the relationship between the number of spots and the total energy the Sun radiates into space? As more information becomes available, the answers to these and similar questions evolve and give rise to new theories about the workings of the Sun, hopefully in the direction of a better match with reality.

We do not expect and never get a final answer to our questions. To use a popular term: there is no closure. Results we can reasonably consider as correct in the light of current scientific beliefs are impressive, but the investigation must go on. In this book, I unashamedly present a history in which tit-bits and larger chunks are continually added to a body of knowledge. Of course, we shall encounter plenty of controversy and several times we shall be able to witness previously accepted theories and individuals' dearly held beliefs totally shattered. But in spite of all the twists and turns, progress is clearly in evidence.

Our story unfolds in time, and is still going on and still needs historical research, because sky watchers' jottings, made hundreds of years ago, still yield significant data. In this respect astronomy is different from some of the other sciences. We tend to think of the sciences as being in the here and now, not much concerned with their own history. Chemists have no need to know how an element was discovered as long as they know the properties of the element and its interactions with others. Similarly, however interesting the story of the discovery and development of penicillin may be, it does not have any bearing on pharmaceutical research today. By contrast, the study of old records has thrown light on past activities of the Sun from which important conclusions can be drawn. We like to think of the Sun as permanent and reliable, though occasionally (in England much of the time) obscured by clouds, but in reality it is a changing star and its output has varied significantly over time. By comparing the Sun to other, similar stars we can chart its history in the longer term and predict its future.

Following the Sun's changes during human history also has practical applications. Sudden changes in the Sun's physiognomy can bring about major calamities on Earth. Particles ejected from the Sun carrying magnetic fields can disrupt telecommunication and blow up transformers in the electricity supply grid. Variations in the Sun's output over a number of years might play a considerable role in climate change and in particular it might contribute to global warming. There is no consensus on what causes this warming trend which extends over the whole of the Earth. Indeed, for a long time there was not even consensus on whether global warming existed at all — and if it did, by how much the Earth has warmed up during the last decades.

Opinions on these vital questions keep changing with disturbing regularity and are often influenced by economic and political considerations. The study of sunspots may provide important and objective clues. This is one of the reasons why old sunspot observations are studied not only by historians of astronomy but also by palaeontologists, meteorologists, climatologists and, by the latest arrivals on the scene, the environmentalists. The old records help us make sense of the Sun's and the Earth's past and at the same time, we hope, they will help us make new and valid predictions for the future.

⌒ 2 ⌒

Do Gods have Blemishes?

All around the globe, the remains of temples built and dedicated to various Sun-gods bear witness to Sun worship from ancient times to our days. Fear of the Sun's disappearance and with it the end of all life has prompted prayers and rituals ranging from sun-dances to sacrifices. Religious practices frequently demand the first prayer of the day at sunrise with the devotee turning to the East. Practitioners of the ancient art of yoga start the morning with a Sun salutation exercise routine, while modern beach resorts cater for sun worship of a different kind.*

Ancient Egyptians venerated a panoply of gods and amongst them were several powerful Sun-gods. One of the earliest was Horus. As the rising Sun, driving away darkness and avenging the murder of his father Osiris, Horus was always represented by a falcon. The Sun-god Ra (or Re) whose principal seat was Heliopolis (Sun City) also had a head of a falcon, or a hawk, but with the addition of a circular sun disc above it. Ra was the most venerated god during the 5th and the 6th dynasties (about 2,500–2,200 BC) and was supposed to be the ancestor of the pharaohs. In Thebes the chief god was human-headed Amun (or Amon, or Ammun). Amun was temporarily toppled when the pharaoh Amenhotep IV (father-in law of Tutankhamen) came to the throne in about 1375 BC. He instituted a religious reformation and placed Aten instead of Amun at the highest position. Aten was the word for the actual, physical Sun and Amenhotep IV, who changed his own name to Akhenaten, decreed that no other deities were to be worshipped. Aten was not allowed to be represented by human or animal forms, only by the disc of the Sun from which rays spread to the Earth below, ending in a hand holding the Nile key symbolizing the Sun's gift of life to the

The Sun-god Horus.

* It would be interesting to hear what sociologists and cultural historians have to say about the recent warnings to shun the Sun.

Above: *Apollo.*

Below: *Amaterasu, the Japanese Sun-goddess.*

world. During Akhenaten's reign Amun and his worshippers were actively persecuted and their temples and their statues were destroyed. But after Akhenaten's death the Egyptians soon returned to the older, polytheistic form of religion and Amun was reinstated. Powerful Amun-Ra was the result of its merger with Ra.

Ancient Greeks, who also had several gods, believed that the flat Earth was surrounded by a river into which the Sun dipped every evening to rise again refreshed the next morning. Then the mythological Sun-god Helios drove his chariot with four horses across the sky. From the fifth century BC onwards, Helios was identified with beautiful young Apollo. Ancient cultures freely borrowed gods from each other, thus Romans carried on with the worship of Apollo, giving him the epithet Phoebus. Similarly Baal, venerated as god of the heavens by the Hittites in the time of Ramses II, later became the Sun-god of the Phoenicians and we know from the Bible that Baal had followers among the Israelites as well. The centre of its cult was in Baalbek, afterwards renamed Heliopolis when the Greeks identified him with Helios.

Myths and legends have surrounded the solar deity in all cultures as life giver and life sustainer. As gods or goddesses the Sun and the Moon were often endowed with human personalities complete with all the frailties humans are prone to. The central figure of the Japanese Shinto religion is the Sun-goddess Amaterasu. One of the stories about Amaterasu recounts that when her brother, the storm god upset her, she retired to a cave leaving the world to grow cold and barren. Amaterasu kept on sulking for a long time and emerged only when other gods hung a mirror at the entrance of the cave in which she could admire herself in her

18

*Ceramic sun face
from New Mexico.*

own reflected light. This round mirror, from then on symbolizing the Sun, was given for safe keeping to the first (mythical) emperor, ancestor of all later emperors.

In some of the stories woven by American Indians their predecessors asked Coyote to rescue the Sun from a cave; in others the Sun is captured and compelled to describe a certain path every day. In yet another version a spider brings up the Sun in its web every day from a cave, or from the underworld, where it spends the night. Inca rulers also claimed to be descended from the Sun just like the Japanese emperors and Egyptian pharaohs. This conferred legitimate divine power on them.

In Chinese mythology the Sun-goddess once gave birth to ten suns. They all lived in a mulberry tree. Every morning the goddess bathed one of the suns and let him fly on the back of a crow. At first the crow took the Sun to the top of the tree and from there flew with it into space. In the evening the Sun would return to the tree. One day all ten suns escaped and flew up into the sky. It was getting far too hot and dry on the Earth and people kept begging them to come down. The suns wouldn't listen and just stayed up in the sky. The situation was getting desperate until Yi, a crack archer managed to shoot down nine of the suns. Ever since this incident there has been only one single Sun, with a three-legged crow living in it. Occasionally the crow gets hungry and nibbles away part of the Sun. Appropriately, in China the traditional symbol of the Sun was a red disc with a crow inside it.

No written records have survived from the prehistoric builders of Stonehenge and similar stone circles, but it has long been suspected that at least one of their functions between 3000 and 1000 BC was intimately connected with solar observation or solar worship. The alignment of the stones with sunrise on the day of the summer solstice and with sunset on the day of the winter solstice is unlikely to be a coincidence.

Did ancient Sun worshippers notice any blemishes on their object of veneration? Here a dose of scepticism comes in handy. Often the Sun has black markings on its face on paintings and carvings (petroglyphs) found in caves or on rocks. We could interpret them as spots but it is far more likely that they were features on the human face of the personified Sun people worshipped. In an Egyptian hymn the Sun-god Ra enjoys the smell of perfumes with his nostrils. Some imaginative souls took this as a reference to sunspots, all the more so because sunspots often come in pairs just like nostrils do. The trouble is that although Ra was the Sun-god, he was not the Sun itself. Sometimes the Sun was taken to be one of Ra's eyes but eyes rarely have nostrils. A supposedly pre-Columbian Aztec myth in which a Sun-god with pockmarks on its face created the world sounds a little more convincing, that is, until we realize that smallpox was not known in the Americas before the Europeans imported it in the sixteenth century.

A story originating from the area of the Zambezi in Africa is more believable. Consumed with jealousy, the Moon throws mud onto the face of the Sun whenever possible. Luckily this does not happen very often because the Sun is watchful. But every ten years the Sun loses concentration and is bespattered with mud.

THE ENIGMA OF SUNSPOTS

Since it is now well established that about every eleven years the number of spots reaches a maximum, could sightings carried over several cycles have given rise to this story?

Written observations are more reliable than orally transmitted myths and legends. Ancient Babylonians kept their official records on clay tablets. Those who are familiar with cuneiform, or wedge, writing can read about the Sun and the Moon and about their eclipses, also about the stars and about comets.

The records show that Babylonians were precise and accurate observers and took their astronomy seriously. One tablet, written more than three thousand years ago and now kept in the British Museum notes that on the first of the month called Nisan (the Babylonian New Year), the Sun appeared to have spots. This is probably the first written report on sunspots and amazed an expert on Babylonian literature. He exclaimed: 'Who would have thought of looking for a notice of sun-spots in the clay tablets of ancient Babylon?'[1] Babylonians kept meticulous records of the weather as well and were convinced that the same weather pattern returned after a cycle of twelve years. Did the Babylonians know about the sunspot cycle?

In the fourth century BC a Chinese astronomer claimed to have seen an eclipse start up in the middle of the Sun as the yin of the Moon overpowered the yang of the Sun from the middle of the disc. He could not possibly have meant a real solar eclipse; as an astronomer he must have known that eclipses always begin at the edge and not in the middle. Could it have been a large sunspot?

In traditional Chinese culture the heavens were supposed to have a sympathetic influence on events here on Earth and it was the task of the astronomer to find out what this influence might be. Future plans had to be made accordingly and reputations rested on correct predictions. So it is quite likely that spots on the

Tablet of Shamash Babylonian, early ninth century BC. From Sippar, southern Iraq (British Museum). The restoration of the Sun-god's image and temple. This stone tablet shows Shamash, the Sun-god, seated under an awning and holding the rod and ring, symbols of divine authority. The symbols of the Sun, Moon and Venus are above him with another large Sun symbol supported by two divine attendants. On the left is the Babylonian king Nabu-apla-iddina between two interceding deities.

Sun were seen many times and were transformed into the mythological crow, the one that lives in the Sun.

Unfortunately many ancient Chinese books were lost through fire and warfare and many more were destroyed in deliberate book burnings. As a result the information we gain from later compilations is sporadic. Furthermore, the surviving records of sunspot observations are rather sketchy and random compared to the regular documentation the Chinese kept of the rest of the sky. A more complete history of sunspots would be useful in charting the changes that we now know undoubtedly occur in the Sun.

Diligent researchers found more than 150 sightings of what we think were sunspots recorded in China, Japan and Korea before AD 1610.[2] All of these are naked eye observations since telescopes were unknown at that time. This leads us to assume that they are only of exceptionally large spots or large groups of smaller spots. Small ones could not have been seen without the aid of a telescope.

The first written record that we can safely assume talks about the dark blotches that appear sometimes on the Sun dates from 165 BC. It notes: 'The character of Wang [King] appeared in the Sun.'[3] The Chinese used several other similarly poetic descriptions and tell us about a 'flying magpie,' a 'flying swallow,' 'dark vapour,' a 'peach' or a 'plum' in the Sun. A book from AD 695 mentions that wanderers came to China from a foreign country bringing a 'precious stone suitable for observation of the Sun.' This object was probably a piece of glass, over a foot in circumference. When looking through it at the Sun, a large floating palace could be made out. Was this yet another poetic description of a sunspot?

Far Eastern records hugely outnumber European ones and the few European references to dark patches on the Sun are found in historical chronicles, not in astronomical records. Did Europeans not look into the Sun, were they saving their eyesight or, were they reluctant to put into writing what they saw?

Some ancient meteorological descriptions could be interpreted as referring to spots. Aristotle's best pupil, Theophrastus,

THE ENIGMA OF SUNSPOTS

a botanist who also studied meteorology, said that rain can be expected if the Sun sets in cloud, if the sky is red, and 'again, if the Sun when it rises has a black mark.'[4]

Two hundred years later the Roman poet, Virgil wrote in one of his agricultural poems that: '[the Sun] when hidden in cloud, he has chequered with spots his early dawn, and is shrunk back in the centre of his disc, beware of showers' and at setting: 'Oft we see fitful hues flit over his face: a dark one threatens rain; a fiery east wind; but if the spots begin to mingle with a glowing fire, then shall you see all nature rioting with wind and storm clouds alike.'[5] Of course, they may have been talking not about spots on the Sun but about terrestrial cloud formations in front of it. On the other hand, spots seen and taken for granted would not be surprising since forecasting the weather is important for farmers and every bit of the sky has always been scoured for tell-tale signs of what can be expected.

Note that both Theophrastus and Virgil couple spots with impending rain. Of course, Virgil may have simply copied from the earlier writer. It is also possible that folk tradition about sunspots in Ancient Rome was similar to that in Ancient Greece. Many centuries later trying to find a correlation not only between sunspots and the weather but between sunspots and just about everything else here on Earth would come to be a favourite occupation both of cranks and of scientists. Is there really a connection or, is a presumed connection pure coincidence? Are we dealing here with astrological fantasy which cannot be explained or, with possible physical influence? Later we shall be able to give at least a partial answer to these questions.

A thousand years separate these ancient meteorological observations and the next known surviving European records about spots on the Sun. We find a ninth-century reference in a biography of Charlemagne, King of the Franks and Emperor. Written by Einhard, secretary, friend and adviser to Charlemagne, it is not exactly factual but contains many imaginative and improvized details. Einhard took the then well known book *Life of the Caesars*, written by the Roman historian Suetonius, as a model and had to fit Charlemagne's life and character to that of the Roman emperors. But he was not entirely making things up when he said:

> Many portents marked the approach of Charlemagne's death, so that not only other people but he himself could know that it was near. In all three of the last years of his life there occurred repeatedly eclipses both of the Sun and the moon; and a black-coloured spot was to be seen on the Sun for seven days at a stretch.[6]

He was not making up things but neither was he particularly careful about dating them. Charlemagne died in AD 814. The portents as described, in the last three years of his life, should have appeared between 811 and 814. True, there were eclipses but not all occurred in those three years. On the other hand, Einhard's text could be translated as: 'For three years at the end of his life' and this would allow a wider margin. In another passage Einhard mentions the fall of a portico as a portent but that fell only in 817, well after Charlemagne's death! There is no doubt though, that a large black spot was seen in the Sun in 807 because records from a German monastery confirm the event.

In the Arab world, where astronomy was based on Greek tradition just as in Europe, and where it was well advanced, we find similarly few mentions of sunspots. Al-Kindi saw one in AD 840 and thought it was the planet Mercury passing in front of the Sun. He feared that it might cause 'various calamities.' In the twelfth century Ibn Rushd (Averroes) and Ibn Bajja (Avempace) likewise interpreted the spots they saw as planets.

In one of the most important English chronicles, that of John of Worcester, the first sunspot drawing appears: two large spots seen on December 8, 1128. (see p.113).

> There appeared from the morning right up to the evening two black spheres against the Sun. The first was in the upper part and large, the second in the lower and small, each was directly opposite the other as this diagram shows.

The official Korean chronicle of the time gives an account of a prominent red aurora seen five days afterwards, on December 13.

This near coincidence makes it very likely that the account and our interpretation of it are both correct because, as we shall see, there is a valid scientific explanation of why spectacular auroras often occur a few days after a large display of sunspots.

A note from Bohemia in 1139 about a fissure, and two fourteenth century reports from Russia complete the European sightings before 1590. In that year James Welsh, master of the ship *Richard of Arundel*, registered events encountered during the ship's voyage to Benin.

> On the 7 [of December] at the going downe of the Sunne, we saw a great blacke spot in the Sunne, and on the 8. Day, both at rising and setting, we saw the like, which spot to our seeming was about the size of a shilling.[7]

And on December 16: 'This night we saw another spot in the Sunne at his going downe.' The sailors complained of a lot of rain and wind they had to contend with during their fifteen-month-long voyage. For once folk tradition and the weather were in agreement. But the spot was not an unlucky sign because they managed to bring home 150 elephant tusks, 589 sacks of pepper and 32 barrels of palm oil. They also 'did kill a great store of small Dolphines and many other good fishes ... which was a very great refreshing unto us.'[8]

Sailors used to navigate by the stars and by finding how high the Sun was above the horizon. It is no wonder then that another reported sighting is also from aboard ship. In May 1609, when they were near the Faroe Islands, the crew of Henry Hudson's *Half Moon* reported the Sun having a 'slake,' this being a northern expression for mud or slime.

In 1605 the poet and painter Raffael Gualterotti claimed in his book *Discorso sopra l'apparizione de la nuova stella* (*Discourse on the appearance of the new star*) that for several days he followed the movements of spots on the body of the Sun. He was certain that it could not be the planet Venus because that would have been much smaller. Gualterotti dreamed up a neat explanation for the spots he saw: the conjunction of the planets Mars and Saturn attracted vapours and exhalations which were then

purified, rarified and drawn to the Sun to become sunspots. The spots disappear after a while because they are shot back into space and turn into new stars. Gualterotti's poetic nature had nothing to do with this strange story, 'exhalations' were commonly used as explanatory devices. We use them too, only give them different names, more in keeping with our current theories.

This handful is nearly the sum total of the early sightings recorded in Europe. Historians have suggested a reason for the scarcity of European and Arab observations, especially when we compare them with the much greater number of Far Eastern ones. The reason, they think, can be found in the picture of the world people had in their minds which prevented them from seeing the imperfections in the Sun. The Greek philosopher Aristotle is usually blamed for this picture. According to his teaching the Moon and anything further away is perfect and unchangeable. We shall see in the next chapter why he thought this must be so.

Even the great astronomer Johannes Kepler fell foul of the idea embedded in his mind that the Sun must be perfect. In 1607 he projected an image of the Sun through a crack in the roof onto a piece of paper and saw a dark fleck in the middle. He instantly believed it to be the planet Mercury. Having seen such an important and exciting event, Kepler felt the urge to tell the whole wide world about it. He immediately sent word to his employer, the Emperor Rudolph II, then he rushed out into the street and urged the passers-by to quickly repeat his observation. Later, in a more relaxed mood, he sat down and composed a short book about Mercury seen in the Sun. But what he saw was not the planet Mercury, it was a sunspot. To his credit, Kepler later admitted his mistake, in print.

Schematic representation of Kepler's camera obscura. The same plate is bound into some copies of the book by Fabricius, amended to read "Conclave Osteel Fris" but with the sign of Mercury unchanged.

∼ 3 ∽

The World According to Aristotle

In the not so distant, but dim past, in my childhood, and definitely before the advent of computer games, a popular children's game was 'famous people.' A letter was chosen by sticking a pin into a page of a book and in one minute flat we each had to compile a list of famous people whose name began with that letter. The compiler of the list with the greatest number of names nobody else could think of was the winner. When the letter 'A' came up, the list was certain to contain many Ancient Greek names and Aristotle invariably topped the page. Given the same task Aristotle's name might still be the first to come into our minds yet most of us know precious little about him.

European culture lost sight of Greek scientific achievements after the decline of Rome, only to rediscover them from Arab sources in the thirteenth century. From that time onwards the picture of the world as constructed by Aristotle reigned supreme until the sixteenth century. By now it has been so thoroughly overturned and surpassed that schools and universities do not consider it worth teaching, except to historians of science and to budding philosophers. What is not widely known is that the discovery of sunspots and their regular, systematic observation played a major role in demolishing the Aristotelian picture of the world.

Aristotle was born into a well-to-do family in 384 BC in Stagira in northern Greece. He studied at Plato's Academy in Athens and then as young men are wont to do, he spent some years travelling abroad in search of adventure and knowledge. Afterwards, at the request of Philip, King of Macedonia, he became tutor to the heir of the throne, whom we know as Alexander the Great. When Alexander grew up, Aristotle returned to Athens and founded his own school, the Lyceum. Teachers and students often walked around while lecturing and studying and from the Greek expression of this, Aristotle's pupils were called the *peripatetics*. Later writers referred to the *peripatetics* when they meant Aristotle and his followers. Politics disguised as morals compelled Aristotle to flee Athens in the last year of his life. He died in 322 BC.

Aristotle worked on philosophy, politics, literary criticism, logic, ethics, biology, cosmology — there is hardly a subject in an encyclopaedia he has not contributed to. He was a fine biologist, a brilliant thinker and an excellent observer. He was however, not born to be an experimenter. His observations, coupled with logical deductions directed by common sense led him to utter opinions that we now consider false.

Aristotle built on, and expanded, the theories of some of the earlier Greek philosophers and astronomers. Everything in the universe was supposed to be made up of four elements: earth, water, air and fire. To these Aristotle added a fifth element: aether. In the Greek astronomical tradition the heavenly bodies were supposed to be carried around the Earth in circles, on imaginary fixed spheres. These spheres, originally introduced only as aids to calculations took on physical reality in Aristotle's mind and became crystal spheres.

Aristotle observed the movements of bodies on Earth and found that if nothing was holding them up or forcing them to move, they would either fall or rise. An apple falls off the tree when the branch no longer holds it, on the other hand smoke from a fire tends to rise upwards. He concluded that the four elements on Earth each have their allotted places they are striving to reach. Heavy bodies, made up from much Earth and water, fall downwards like the apple, but air and fire strive to reach their place higher up.

Of course, Aristotle was well aware that a stone tied to a string and twirled around will move in a circle, but he also knew that this was because we force it to do so. As soon as we let go, the stone will start falling and he reasoned that its wanting to fall would eventually overcome the remnants of our force. Smoke would also move upwards in a straight line if it wasn't for the air pushing it around. From these and other similar observations Aristotle deduced that terrestrial bodies have the intrinsic property of moving in straight lines either up or down, either towards the centre of the Earth (the centre of the Universe for him) or away from it. But when Aristotle looked up at the sky, he found that a different law was in operation there. The stars did not fall towards the Earth but kept moving around it. From this observation he deduced that the natural movement of the heavenly bodies is circular.

Earthly objects had yet another property not shared by the heavens: they were all subject to changes. Living things are constantly born, grow in size, change in all manners of ways, grow older and finally die. Inanimate objects change with time too: for instance fire extinguishes, a river changes its course and eventually even rocks erode. But the stars seemed to be unchanging. Aristotle was a philosopher and he tended to generalize. The outcome of his generalization was the following. Objects that are closer to the Earth than the Moon are subject to generation and destruction and their natural movement is in a straight line. The Moon and all other objects that are further away are 'ungenerated and indestructible and exempt from increase and alteration'[1] and their natural movement is circular. Obviously, they must be made from a different material.

Aristotle argued further, that the Sun and the stars are not hot. He said: 'The warmth and light which proceed from them are caused by the friction set up in the air by their motion. Movement tends to create fire in wood, stone and iron ... an example is that of missiles, which as they move are themselves fired so strongly that leaden balls are melted.'[2] Here again, Aristotle relied on his observations: heat is created by friction when bodies are rubbed against each other. But his assumptions, that the Sun orbits around the Earth and that an atmosphere fills the whole of the universe, were wrong.

There were, of course, philosophers who held views different from Aristotle's about the universe. In an earlier age Anaxagoras luckily escaped with his life when he was prosecuted for impiety because he taught that the Sun was a great burning mass, maybe

Opposite:
Ptolemy with the muse of Astronomy

THE ENIGMA OF SUNSPOTS

In the illustration: *Ptolemeus*, *Astronomia*

Copernicus and the Copernican system.

as large as the Peloponnese. Aristarchus who lived after Aristotle and who believed that the Earth moved around the Sun was not similarly charged but neither was he successful in making his teaching accepted.

In the first century the Alexandrian astronomer Claudius Ptolomaeus, or Ptolemy as he is called in the English speaking world, presented the views of the Ancient Greeks about the universe in a practical, mathematical form in his book the *Almagest* (the *Greatest*). In Ptolemy's world the Earth sits in the centre, the stars are fixed on nested spheres around it revolving in circles, while the planets revolve in circles on circles. It is very likely that Hipparchus worked out the details originally a couple of hundred years earlier, but such is the injustice of fame that we call it the 'Ptolemaic system.'

After their rediscovery, the European establishment tried rigidly to uphold Aristotle's opinions and the Ptolemaic world system for over two hundred years, even in the face of accumulating experimental evidence. The Catholic Church would declare those who questioned Aristotle's statements as heretics and the Inquisition was more than ready to put them on trial. The official view at the end of the sixteenth century was still that of a perfect and unchanging sky above us and indeed this could have been the reason for the paucity of the records of European sunspot observations. If people were convinced that the Sun was immaculate then it could not have black spots on its face. If they were convinced that no changes ever took place on the Sun then no spots could appear and disappear on its disc. In other words, it is not easy to override a mental state once we are conditioned to it and if we believe that something does not exist then the

THE ENIGMA OF SUNSPOTS

chances are, that we will not see it. It followed that any black spot seen on the Sun could exist only either in the imagination of the beholder, or was a floater in the eye (one of those little dark specks that occasionally float around in our visual field) or due to some atmospheric effect. In any of these cases the observation should be disregarded. In addition, few had the temerity to openly contradict the teachings of the Church, especially since deviant beliefs could lead to imprisonment or burning at the stake.

By the late sixteenth century big changes were already afoot in Europe. At the beginning of the century the Polish astronomer Nicolas Copernicus had developed the idea that the Earth orbits around the Sun together with the other planets. He stripped the Earth from its privileged position and although he did not exactly uphold the Sun's divinity, he gave it a more elevated and central role.

Other advances swiftly followed. In the middle of the century the talented and industrious Dane, Tycho Brahe discovered a hitherto unseen new star that suddenly burst on the sky. He called it a 'nova.' This event fundamentally contradicted Aristotle's theory of the eternally unchanging firmament.*

Tycho set up an observatory on an island between Sweden and Denmark. He furnished it with the best instruments of the time. They were his pride and joy, expensive, large and very precise. Using these instruments to measure angles Tycho and his assistants charted the motion of the Sun and the planets in order to prepare astronomical tables. After Tycho's death, his most famous assistant Johannes Kepler finished the work of the Danish astronomer. The precise astronomical tables Kepler

Tycho Brahe.

* In fact, 'Tycho's star' as it became known, and the other novas and supernovas are not really new stars, but previously unobserved faint stars that suddenly explode and turn very bright.

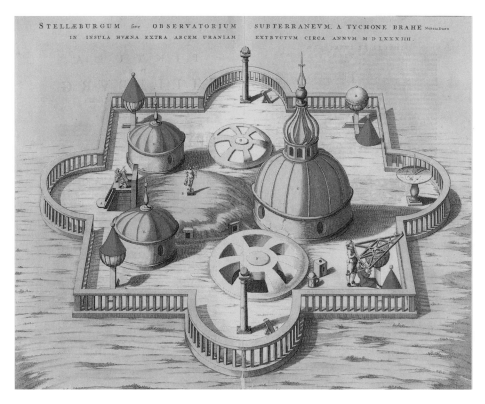

STELLÆBURGUM sive OBSERVATORIUM SUBTERRANEVM. A TYCHONE BRAHE Nobili Dano

IN INSULA HVÆNA EXTRA ARCEM URANIAM EXTRVCTVM CIRCA ANNVM M D LXXX IIII.

Artist's impression of Tycho's observatory. compiled helped him to develop a theory that made another dent in Aristotelian cosmology. Kepler showed that the orbits of the planets are not circles on circles at all, but ellipses. This was a final death blow both to the idea that the natural motion of the celestial bodies was circular and to the existence of fixed crystal spheres.

~ 4 ~

A New Instrument Opens New Worlds

Tycho still conducted his observations with the naked eye, but in 1608 an invention came to the market in Holland that was to change once and for all the field of astronomy. The invention consisted of fixing two lenses into the two ends of a hollow tube. At first people referred to it as the spy-glass. We call it the telescope.

Spectacle making was an established trade by then and standard lenses were easily available. So there can be no doubt that some people experimented with arrangements of several lenses. This may have been the reason why the two men who independently claimed to have invented the new instrument, Jacob Metius and Hans Lipperhey (or Lippershey), were both unsuccessful in their patent applications. A third major contender was Sacharias Jansen and historians have still not decided to which of the three the honour is due.

Two lenses in a tube: it sounds simple enough. But the truth is that it was, and still is, far from simple. It is not simple to grind and polish glass lenses with a suitable and constant curvature and, as far as possible, without any external and internal blemishes. The correct distance between the lenses had also to be found. The experience of the Bavarian astronomer Simon Marius shows that this was not an easy matter. Marius' patron Johann Fuchs of Bimbach saw a spy-glass at the Frankfurt Fair in 1608 but considered the asking price too much for such a flimsy instrument. Returning home he discussed the matter with Marius and the two decided to construct a telescope themselves. In Marius' words:

Optical grinding apparatus.

We afterwards took glasses out of common spectacles, a concave and a convex, and arranged them one behind the other at convenient distance, and to a certain extent ascertained the truth of the matter [i.e. that it produced an enlarged image]. But as the convexity of the magnifying glass was too great, he made a correct mould in plaster of the convex glass, and sent it to Nuremberg to the makers of ordinary spectacles that they might prepare glasses like it; but it was no good, as they had no suitable tools ... If we had been acquainted with the method of polishing glasses, we should have produced excellent spy-glasses after our return from Frankfurt.[1]

Standard spectacle lenses such as the tradesmen could produce were not good enough and since neither Marius, nor his patron could grind lenses, they had to wait until the summer of 1609 for a reasonable instrument to reach them from Belgium.

Galileo Galilei heard about the invention when he was staying in Venice in May 1609. It was exciting news. After his return to Padua where he was professor of mathematics, he wasted no time in building a telescope for himself. Into a lead tube:

I fitted two glasses, both plane on one side, while on the other side one was spherically convex, and the other concave. Then applying my eye to the concave glass I saw objects satisfactorily large and close. Indeed they appeared three times closer and nine times larger than when observed with natural vision only.[2]

This means that Galileo's first telescope magnified 3 times (the area of the picture was $3 \times 3 = 9$ times the original area). At first he probably bought the lenses from a spectacle maker. But Galileo was able to grind and polish his own lenses and could figure out the distance that was needed between the two whatever

THE ENIGMA OF SUNSPOTS

curvature he chose. He kept improving his telescopes and within a year he achieved a magnification of twenty and possibly even thirty times. He also added an aperture stop that reduced the problems caused by optical imperfections. Galileo's telescopes and the demonstrations he gave with them in Venice and in Rome were much admired by all the influential men there. He made several instruments to give away as presents, partly in order to further his career and partly so that his wonderful discoveries could be verified. As a spin-off he got tenure with increased salary at Padua, where he was professor at the time. This did not prevent him from moving to Florence in that same year.

Later Kepler suggested making telescopes with two convex lenses. Such telescopes have greater magnification but produce an inverted image. This is a drawback for generals watching their troops on a battlefield but not a problem for astronomers. Sunspot watchers hailed it as a positive advantage because the Keplerian telescope rectified the inverted image they obtained when projecting the Sun onto a piece of paper. Later the reflecting telescope introduced by Newton, using mirrors instead of lenses, gained popularity.

Within a few years anyone who was interested in astronomy and who could afford it acquired a telescope. Those who could not afford it tried to make one. Some were successful like Galileo, others less so like Marius. These instruments, although of rather poor quality and low magnification compared to modern telescopes, nevertheless enabled their owners to look at the sky with new eyes. New discoveries came to light, some were real, a few were imagined. It must have been an exciting time to live.

～ 5 ⚬

A Hasty Publication

Young Johann Fabricius, son of the astronomer/astrologer David Fabricius, rushed into publication soon after he noticed something unusual on the Sun, early in the year 1611. Father and son conducted their observations together but it was Johann who first saw the spots and realized that he was seeing something strange, something he had never encountered in the course of his studies.

Fabricius senior was a Protestant vicar in a village in Ostfriesland (he and his son are sometimes referred to as Dutch, sometimes as German; actually Friesland is in Holland but Ostfriesland is now a province of Germany), on the estates of Count Enno. The count encouraged and financially supported both the astronomical and the astrological activities of his vicar. David Fabricius befriended several astronomers and in 1601 travelled, at the count's expense, to Prague to meet Tycho who was living there at the time. Tycho died soon afterwards, but his Prague visit gained Fabricius a valuable correspondent for a time in the person of Kepler.

David Fabricius' claim to fame rests on his naked eye discovery of the first variable star later named *Mira Ceti* or, a 'miracle in the constellation of the whale.' He saw the star first in 1596 and measured its exact location. He knew it was not a comet because it had no tail and at first he thought that it was a new star, a *nova*, like the one Tycho discovered in 1572. But this new star disappeared from view. To Fabricius' surprise it reappeared again in 1609 in exactly the same position as before. From this Fabricius deduced that it must have been in the sky all the time, only varying in brightness. Sometimes it was so faint that he could not see it, then it gathered strength and shone brightly

THE ENIGMA OF SUNSPOTS

again. Needless to say, this was another nail in the coffin of Aristotle's theory.

David Fabricius, like many astronomers of his time was also a practising astrologer. We don't know how much of his stargazing can be ascribed to a wish to find out about the wonders of the world and how much to wanting to find a connection between the stars and human affairs. Certainly, he explained the appearance of comets and new stars as signs of God's anger, while the horoscopes he compiled for his patrons must have been a lucrative sideline. How strong his belief in horoscopes was is questionable. He had a premonition of his death and set up a horoscope accordingly for himself yet did not try to escape. In 1617 David Fabricius was murdered by one of his parishioners, a suspected thief. This man hit Fabricius on the head with a spade to forestall being denounced by him from the pulpit.

David's son, Johann, was a restless youth. He moved from one university to another and regularly changed the subjects he was studying. Such behaviour was more acceptable in those days than it would be today. As long as a parent or a patron was willing to pay his expenses, a young man was free to roam the world. Johann started out as a medical student in 1604 but the only qualification he ever received was Magister Philosophiae in 1611. As luck would have it, one of the universities where he spent some time was Leyden and there he could buy specimens of the newest invention, the telescope.

As a budding astrologer young Fabricius fervently wished to make a great discovery. He decided to concentrate his attention on the Sun because it was the least studied celestial object and because he was already interested in the connection between the Sun and the weather. Early in the morning of March 9, 1611, alerted to an irregularity his father saw near the Sun's edge, Johann set up one of his instruments in the loft of his parents' home. His attention quickly shifted from the edge of the Sun as he peered through the telescope and was amazed to see a large black spot on the Sun. It might be just a small cloud, he thought. But the spot moved with the Sun across the sky while a cloud would have disappeared from view. He looked through

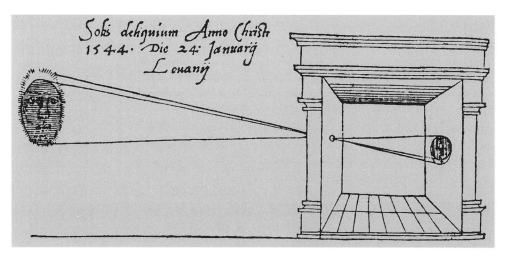

Gemma-Frisius' illustration of a camera obscura. This plate is thought to be the first illustration of a camera obscura.

several other telescopes and the spot was there every time. He was still not sure of himself and called down to ask for his father's advice. David Fabricius climbed up to the attic and the experienced sky watcher verified his son's discovery. The two got so excited that in spite of severely aching eyes they watched the Sun until noon.

That night neither father nor son could sleep. Their thoughts, which kept returning to the telescope in the attic room, kept them awake. They wondered whether the spot would be visible the next morning. As soon as the Sun was above the horizon they rushed upstairs and resumed their observation. Sure enough, the spot was there. It was not exactly in the same position as on the previous day, it seemed to have moved. After a while their eyes ached so much from looking into the bright, hot Sun, that they had to give up watching. Then they had a brainwave and tried another method. Closing all the windows and doors, they left only a small opening for the telescope. In that dark room (the *camera obscura*, which means dark room, in Latin) they placed a piece of paper at a distance from the telescope and projected the image of the Sun onto it. This method of observation was much kinder to their eyes than looking straight through the telescope into the Sun.

THE ENIGMA OF SUNSPOTS

The *camera obscura* had been used previously, with or without a lens in the opening, for looking at solar eclipses. When Kepler observed what he believed to be Mercury transiting the disc of the Sun, he used the same arrangement without a telescope. After the invention of the telescope this method became popular for observing sunspots. A miniature *camera obscura*, called a 'pinhole camera,' can be made by piercing a tiny hole into one side of a box. An image will be cast on the opposite side. Galileo was much 'impressed by the courtesy of nature' providing this method whereby 'without any instruments from any little hole through which sunlight passes, there emerges an image of the Sun.'[1]

Fabricius could watch the projected image of the Sun day in day out. The large spot moved slowly from east to west and eventually disappeared while some other, smaller spots appeared at the eastern edge. These also moved towards the west. After ten days or so, the large spot reappeared on the eastern edge. Young Johann Fabricius persevered in his observations and was able to see several more appearances, disappearances, and reappearances. He noted that the spots moved faster when they were in the middle of the disc and slower at the edges. He saw them changing their shape, expanding when they were in the middle of the Sun and getting thinner as they reached its side.

Illustration of a camera obscura. (1657). From Magia Universalis *by Gaspar Schott*

Then he asked himself: where are these objects, are they on the Sun or are they at some distance from it? This was one of the most important questions the first serious observers in the seventeenth century had to confront. They did not all come to the same decision. Some spots disappear in the middle of the disc but others persist, cross the disc, disappear at the opposite edge and reappear on the other side in about two weeks' time. If the spots were on the body of the

Sun then the inevitable conclusion would be that the Sun itself revolves.

In his later account of the observations Fabricius concluded that: 'If outside the Sun, it would not be possible to detect them on following days on the Sun's body, since by its proper motion the Sun bypasses any small cloud or body suspended between us and the Sun.'[2] From the movements of the spots and from their changes of shape Fabricius deduced that the Sun must be a solid, round body that revolves around its own axis. To make the last statement acceptable, he carefully reminds us that this had already been asserted by Giordano Bruno (who was burnt at the stake) and by Kepler, 'a man of profound erudition and admirable judgement.' He adds that spots on the Sun had been seen in previous ages but their nature was incorrectly interpreted. They could not be clouds, nor could they be comets in the making.

Johann Fabricius published a little booklet in a hurry for the 1611 autumn Frankfurt Fair, which was and still is, an important meeting-place not only for peddling books but also for the exchange of information. His *De Maculis in sole observatis* ... [Of blemishes observed in the Sun] was not a success. An influential patron could have helped promote the book and have it more widely distributed but Fabricius did not have such a patron. Because of its simple style and its diffident approach, the book was regarded as no better than a student's essay and did not arouse any interest. Indeed, it is short for a treatise and the ideas are not sufficiently developed. Fabricius argues for instance, that at the margins of the Sun: 'many points fall together in one straight line of vision,' therefore it cannot be as he first thought, that the spots move at a distance around the Sun. Galileo used the same argument as Fabricius, based on perspective vision, but he produced a geometrical proof complete with diagrams and calculations. Fabricius only hinted at the same idea without giving an explicit proof. It is possible that his mathematics was not good enough; it is possible that he was in too much of a hurry to get into print. It is also possible that the inexperienced young man believed that his explanation was clear and sufficient, and that

THE ENIGMA OF SUNSPOTS

what he knew and understood everybody else would also know and understand. He does allow us a glimpse at his thought processes though: continuous observation contradicted his first theory that the spots circled the Sun at a distance and this made him change his mind. (The mind of one of the other discoverers, that of Christoph Scheiner, was not similarly open. Scheiner came to a diametrically opposite conclusion.) It is a historical fact that the learned community has ignored Fabricius' reasoning. Publications on the early history of the discovery of sunspots feature only Galileo's proof that the spots reside on the body of the Sun.

Within a few years, during a bitter and publicly conducted priority fight between Galileo and Scheiner, any hopes of fame Fabricius might have harboured disappeared. In the polemics of the two contenders the name of Fabricius does not crop up. They must have known about his book, probably even read it, but ignored it. A suggestion that religious differences prompted the two Catholics not to take any notice of the Protestant Fabricius is unlikely to be correct. In this author's opinion neglect of Fabricius' work was more likely due to the personal vanity of the protagonists.

~ 6 ~

Two Astronomers
Fight for Priority

Soon after Fabricius published his little treatise, another book appeared in which the pseudonymous author also described sunspots and implicitly claimed to be the first to recognize them. As it later turned out, the author behind the pseudonym was Christoph Scheiner, a Jesuit and at that time Professor of Mathematics at the University in Ingolstadt in Bavaria.

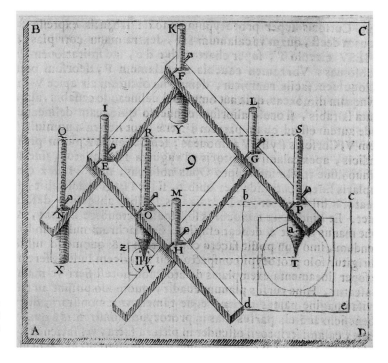

The pantograph, from Scheiner's Pantographice

THE ENIGMA OF SUNSPOTS

Scheiner was a practical man. He invented the pantograph, a drawing instrument consisting of interconnecting linkages which will enlarge, reduce, or reproduce on the same scale, a pattern traced with it. The pantograph and other instruments based on the same principle have been in use ever since. James Watt used it in his beam engine and similar linkages are utilized on electric locomotives to keep the current collector in contact with an overhead wire. Computer aided design has not completely superseded the pantograph as a drawing instrument: versions still on the market are used in industry for shape cutting and engraving. As a toy it gives pleasure to children.

Scheiner owned several telescopes, some he purchased, some he made himself. His best instruments magnified about 25 times. In the spring of 1611, on a slightly foggy morning he climbed up the church tower with one of his students, Johann Baptista Cysat, in order to study the Sun. There is no reason to doubt the report of a well known German scientist that in his diary Cysat gave March 6 as the date of this observation. Nor Cysat's claim that on that morning it was he who first noticed the spots ('drops' he called them), and not the professor. This diary, a two volume manuscript, was seen at the beginning of the nineteenth century. Since then unfortunately, the library in which it was deposited suffered the usual rationalizations, reorganizations, and removals, with the result that the manuscript is now lost. Should this manuscript ever be found, and the date confirmed as March 6, then it would substantiate that Cysat may have seen the spots just prior to Fabricius.

But Cysat was only a student while Scheiner was the professor and the owner of the telescope. Cysat only confided the discovery to his diary while Scheiner proclaimed it to the world and made the study of sunspots his life's work. The fact remains that some time in the spring of 1611 according to Scheiner himself, he saw some blemishes on the Sun and since he was a firm believer in Aristotelian science, did not believe his own eyes. He rubbed those unbelieving eyes, thinking that maybe he had floaters in them. The spots were still there. In any case, his student saw spots as well and it would have been highly unlikely that a strange eye complaint struck both of them at the same time. Scheiner cleaned the lenses of his telescope and resumed his observation through the now spotlessly clean lenses. The sunspots were still there. He tried using several other telescopes in case the lens of the first one had unusually large imperfections in the glass itself. The spots were still there. Maybe the spots were somewhere in the atmosphere. He waited for a while for them to dissipate. The spots did not disappear. In the end Scheiner had to admit that the spots had something to do with the Sun.

Scheiner and Cysat kept watching early in the morning and late in the afternoon. They watched whenever there was some fog or hazy cloud in the sky. A few clear days and strong sunshine made their eyes ache and forced them to stop work but then Cysat had the bright idea of using blue or green coloured glasses to attenuate the light and the heat. They placed the coloured glasses between the object lens and the eyepiece. This was not an original idea; sailors often used them when they had to look into the Sun. It seems however, that although David Fabricius had been known to use coloured filters when watching an eclipse, he did not recommend them to his son for sunspot observation. Johann does not make any mention of them in his book. This was possibly because the quality of coloured glass was still poor at the time and the impurities in it would have made sunspot observations uncertain. When watching a more robust spectacle such as an eclipse, the defects of the glass mattered less. Even sixty years later Robert Hooke's opinion was: 'as to the coloured Glasses, I cannot at all approve of them, because they tinge the Rayes into the same colour, and consequently take off the truth of the appearance, as to Colour; besides, it superinduces a haziness and dimness upon the Figure, so that it doth not appear sharp and distinct.'[1] Hooke advised the application of several reflections to attenuate the rays and this method was used later quite extensively.

Abandoning the use of coloured glasses Scheiner and Cysat hit on the best method then available, and stuck a telescope into the opening of a dark room projecting the image of the Sun onto a piece of paper, just like the Friesian astronomers did. They drew the circular outline of the Sun on the paper and marked the positions of the spots. Then they slowly moved the telescope in tandem with the progression of the Sun so that the image would always fall on the circle. This allowed them to follow the movements of the spots.

Scheiner had a problem. As a Jesuit, he was bound by obedience to his superior who would not let him publish the heresy that the Sun is not perfect. A slight suspicion that Scheiner might be mistaken could also not be ruled out. In that case publication would have made a laughing stock of the Jesuit Order. Had there been an Internet at the time, Scheiner could have circumvented his superior's prohibition by posting something on it anonymously. He did the next best thing and wrote three short letters to Marcus Welser. These were dated from November and December 1611 respectively. Comparison with the Internet is not a facile one. The first scientific journal appeared in France only in 1665 and even for many years afterwards travel and correspondence were the two main routes for the quick transmission of ideas. Letters were used for exchanging information, for posing problems, for asking and answering questions, and also to publicize new discoveries. They were read out to little gatherings, they were often copied and forwarded to other correspondents. Letters were cheaper and quicker than publication, and they were also free of censorship. In the process proper networks developed with one correspondent acting as a hub in the centre. When the Royal Society was established in the 1660s, it also served as a hub since one of its main functions was correspondence. Letters received by the Society that were deemed important were read out at the meetings and then published in the *Philosophical Transactions*. It is worth noting that we are now fast returning, albeit on a different level, to this older method of communication. The difference is that instead of paper we use electronic means.

Welser, the recipient of Scheiner's letters, was a rich and influential magistrate in Augsburg. He also acted as banker to the Jesuits. Welser was on friendly terms with many scientists and was a generous patron. He managed to persuade Scheiner to let him have the letters published under a pseudonym. The pseudonym Scheiner chose was that of the ancient Greek painter Apelles. If Fabricius lacked self-confidence, the professor at Ingolstadt seemed to have plenty of it. Apelles was ranked as the finest painter Greece has ever produced. Did Scheiner fancy himself as Apelles, painting the features of the Sun?

The book appeared with the well known adage: *Apelles latens post tabulam*. This phrase 'Apelles hiding behind the picture' refers to two different stories. According to one of the stories Apelles exhibited a self-portrait so true to life that viewers believed the picture was the living person until he surprised them by stepping out from behind it. The second story might be more in line with Scheiner's intention: Apelles was hiding behind the painting because he wanted to hear honest criticism of his work. The saying 'no day without a line' is also attributed to Apelles and could mean in Scheiner's case that he spent 'no day without watching the Sun.'

In January 1612 Welser sent a copy of the little book hot off the press to Galileo, seeking his opinion of it. Galileo, pleading ill health, answered only in May of that year. Soon afterwards he himself did some serious work on the subject, both theoretically and experimentally and dispatched a second letter to Welser. Meanwhile Scheiner also decided that the matter merited a more accurate evaluation and in that same year produced another little treatise in the form of letters, under the same pseudonym. He called this book *De Maculis Solaribus ... Accuratior Disquisitio* [A More Accurate Treatise on the Spots of the Sun ...].

By then Scheiner had read Galileo's first letter which was translated from Italian into Latin, but his Italian was still not good enough to understand Galileo's second letter written in the vernacular. (Scheiner's Italian improved later when he moved to Rome.) Galileo, on the other hand, did read Scheiner's second publication and still in ignorance of the author's real name, wrote another, third letter to Welser. Galileo's three letters to Welser were also published in book form, though not by Welser, but by one of Galileo's patrons.

What do all those letters contain? Scheiner admits that his interest was only aroused in October, some seven or eight months after his first sighting. We may well ask why this was so and give a speculative answer later on. Scheiner explains why he is certain that the spots are not fancies of his imagination and cites several independent observers who testified to this.

Because at that stage Scheiner was still convinced that the Sun was perfect, as Aristotle declared, he could not admit to any blemishes on it. The Sun was also unalterable according to the dogma, so the very obviously changeable spots could not be on its surface. A solution to his problem was that the spots were none other than small bodies orbiting around the Sun. Scheiner tried to prove that this was possible and that planets could appear as black spots on the face of the Sun. One way of doing this would have been to catch the known planet, Venus, in transit. This would have been the first time ever that somebody consciously observed a planet in front of the Sun. But in vain did Scheiner watch closely at the appointed time; Venus would not appear. As a result he could not prove or disprove anything, except maybe that his observational methods, or his calculations, or the astronomical tables he relied on were defective.

Scheiner carried on watching the spots diligently. He saw the spots gradually change their shape, he saw them move across the Sun's disc and disappear at its edge. He failed to see them reappear and this convinced him, if he needed convincing, that the spots could not be on the body of the Sun but were orbiting around it. It did not seem to occur to Scheiner that maybe he did not recognize the same spot reappearing due to its changed shape and changed position.

Scheiner reasoned that the small planets occasionally congregate and obscure each other. This would explain the spots' irregular shape. Some are so small that when they separate we cannot see them any more. This explains their disappearance in the middle of the disc.

Scheiner was of the opinion that the spots were very dark, darker than the dark areas of the Moon. Finally, it did not escape him that usually an irregular, darker patch is surrounded by a relatively less dark area. Nowadays the darker part is often referred to as the *nucleus* (meaning 'kernel' in Latin), and sometimes the lighter part is called the *umbra* or *penumbra* (meaning 'shadow' and 'partial shadow' respectively in Latin). In Scheiner's days the nucleus was often called the *umbra*. This may sound confusing but in the proper context the meaning is always clear.

In his second set of letters, resulting from many more days spent in watching the spots, Scheiner noted that they moved faster when they were closer to the Sun's equator and slower when they were further away from it. By that time he did believe his eyes more than Aristotle's teaching and accepted the possibility that the Sun may not be uniformly bright. He saw that often the spots were accompanied by very bright areas, much brighter than their surroundings, and these he named *faculae* ('little torches' in Latin).

Galileo's letters, in which he answered Scheiner's publications, plainly demonstrate the difference of the mental powers between the two men. Scheiner was diligent and competent. Galileo was a genius. Scheiner was still thinking within the boundaries of classical concepts and was intellectually bound by his Jesuit affiliation. Galileo was a free spirit. When Scheiner disputed Galileo's idea that the spots were like clouds and asked: 'Who would ever place clouds around the Sun?' Galileo answered simply: 'Anyone who sees the spots and wants to say something probable about their nature.'[2] Scheiner, probably in order to gain the approbation of his superiors, appealed to the authority of classical authors. Galileo was convinced that Aristotle would have changed his theories in the light of new evidence, new experience. As he later put it:

> We do have in our age new events and observations such that if Aristotle were now alive, I have no doubt he would change his opinion. This is easily inferred from his own manner of philosophizing, for when he writes of considering the heavens inalterable, etc. because no new thing is seen or generated there or any old one dissolved, he seems implicitly to let us understand that if he had seen any such event he would have reversed his opinion, and properly preferred the sensible experience to natural reason.[3]

THE ENIGMA OF SUNSPOTS

Here Galileo pays Aristotle the compliment of assuming that they share the same philosophy of science.

Though Galileo contradicted most of Scheiner's assertions, he wrote to Welser: 'I desire to share in your friendship with Apelles, deeming him a person of high intelligence and a lover of truth.'[4] Within a short time these friendly feelings would change into bitter enmity.

In his first letter to Welser, Galileo claims the discovery for himself. The existence of spots on the Sun is not news to him at all. Indeed, he says, he has seen them already eighteen months previously and even showed them to his friends in Rome. The book containing Galileo's letters is thus the third one published within two years claiming priority of discovery.

In his second letter Galileo affirms that the spots are not planets but rather cloud-like structures, and by then he is certain that they are on the body of the Sun, or if not actually on the Sun, at least extremely close to it. He produces a geometrical proof that this must necessarily be so. The proof is based on measurements on how the distance between two spots changes as they move from the edge (or 'limb') towards the middle, and then again towards the other limb. Galileo also draws our attention to the fact that it is only in contrast with the Sun's brilliance that the spots look so dark; actually they are extremely luminous. This last fact had to be argued even two hundred years later when Honoré Flaugergues, a French amateur astronomer and spot watcher, recommended an experiment to those who still thought spots were black. They were to reflect the light of the sky with a tiny mirror on to the Sun and see that it produced a black spot on it.

In the second letter Galileo chides Scheiner for trying to prove that the orbit of Venus is between the Earth and the Sun and could appear as a little spot on the Sun's surface. Did not Scheiner know that he, Galileo, had already proved this by discovering the phases of this planet? Galileo rejects the notion that spots might be lakes or caverns and offers as an alternative, the possibility that they are much like vapours produced when bitumen is dropped onto a red hot surface. With the help of Benedetto

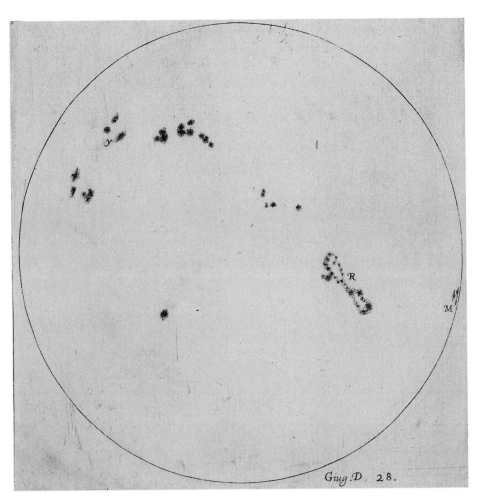

Sunspot drawing from Gailieo's Istoria e dimostrazioni intorno alle macchie solari (1613).

Castelli, one of his pupils, Galileo attached sunspot drawings to his publication. These were far superior to Scheiner's pictures.

Galileo dismissed several of Scheiner's ideas that were correct. He would not accept that the Sun might not be uniformly bright. But Scheiner was right, the surface of the Sun is not uniformly bright, it has a granular nature with lots of darker and lighter areas that are in constant motion. Galileo asserted that the spots all moved with equal speed, Scheiner knew this was not true.

THE ENIGMA OF SUNSPOTS

Galileo at first believed that the Sun rotated around an axis perpendicular to the ecliptic, simply because at the time of year when he first started to watch them they described straight lines in that plane. That he later changed his mind about the Sun's axis was not due to Scheiner's influence but rather to a letter he received from another observer in 1613. To sum up, his observations convinced Galileo that:

> The Sun is absolutely spherical, that it rotates from west to east around its own centre, carries the spots along with it in parallel circles and completes an entire revolution in about one lunar month.[5]

If we stop to think for a moment, it is pretty amazing that so much can be deduced just from watching a few dark patches millions of miles away. Indeed, surprising as it may seem, it is possible to draw many more conclusions. But for the moment Galileo was convinced (speaking of himself in the third person) that:

> In his *Letters* he had looked into and demonstrated everything that human reason could attain in such matters, if not everything that human curiosity might seek and desire, he interrupted for a time his continual observations, being occupied with other studies.[6]

On the other hand Scheiner and some of his Jesuit brothers were not similarly satisfied and kept on pursuing the subject. Another Jesuit disclosed Scheiner's identity in a book published in 1614 and named him as the discoverer of the spots.

A few years later the French Canon Jean Tarde set out to defend the Sun against all those who would put blemishes on it. Although Tarde discussed the sunspots personally with Galileo, he supported Scheiner's original idea that they were small planets and named them 'Bourbon stars.' He explained the changes that spots undergo by the effect of very many very small planets moving with different speeds meeting up occasionally and separating at other times. Tarde's methods of observation must have been deficient because his drawings show the spots as small circles and do not show the penumbra.

The Belgian Jesuit father Carol Malapert could not accept the Sun being spoilt by spots either. He called the presumed little planets 'Austrian stars.' Tarde and Malapert were imitating Galileo who called Jupiter's moons the 'Medici stars' and must have hoped that this would bring them favours from the monarchs of the two countries.

Galileo meanwhile had another bitter priority dispute on his hands, this time with Simon Marius about Jupiter's moons. Being in a bellicose mood he struck out at Scheiner as well, both privately in his many letters to friends and publicly in his books. Although he had discontinued his solar studies, he still insisted on his claim to priority. In a discourse — published under someone else's name but known to be by Galileo — he wrote:

> [people] attempted to make themselves the inventors of views that are really his [i.e. Galileo's], pretending to be Apelles when with poorly coloured and worse designed pictures they have aspired to be artists, though they could not compare in skill with even the most mediocre painters.[7]

He stuck to the story that he had showed the spots to his friends in the Quirinal Gardens in Rome even before Scheiner had set eyes on them. Some years later he wrote:

> The original discoverer and observer of the solar spots (as indeed all the novelties on the skies) was our Lincean Academician [i.e. Galileo]; he discovered them in 1610 while he was still a lecturer of mathematics at the University of Padua. He spoke about them to many people here in Venice, some of whom are yet living, and a year later he showed them to many gentlemen at Rome.[8]

Indeed, some of his friends testified that the demonstration took place in Rome and it is quite possible that one of them mentioned it to Scheiner. Could this be the reason why Scheiner suddenly took up the study of the previously seen but neglected phenomena? But Galileo was not entirely ingenuous in his attack on Scheiner. Even if we accept that he did not know about Fabricius, he must have been fully aware that at least one other person had recognized the existence of sunspots before him and tried to give an explanation. This was Gualterotti. It is unlikely that Galileo was unaware of Gualterotti's observations, since the two corresponded and even knew each other personally.

As to Scheiner, he was deeply hurt by Galileo's attacks. One day browsing in a bookshop he inadvertently overheard somebody praising Galileo's book. An eyewitness described the scene that followed. Scheiner became apoplectic: 'he turned livid and shook so badly that the bookseller' and the other customers were shocked.[9] And he hit back at Galileo.

Denying the Ptolemaic world system was considered heresy and the Inquisition tried Galileo for it. There might be some truth to the accusation that in seeking revenge Scheiner contributed behind the scenes in bringing Galileo to trial, since we know that in his letters to other correspondents he repeatedly denounced Galileo. If so, it must have been vengeance for purely personal reasons because Scheiner slowly changed his opinion about the immobility of the Earth, and by the time of Galileo's trial he regarded the issue not as a matter of faith but something permissible to be questioned. So it is unlikely that he acted, indeed if he did act, from religious motivation.

Scheiner's *magnum opus,* the

One of many drawings from Scheiner's Rosa Ursina *(1630). Note that it covers two weeks of observations following the spots' change of shape and position.*

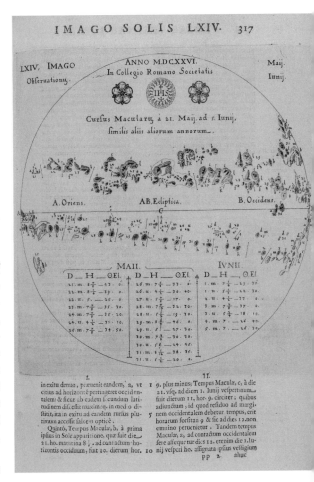

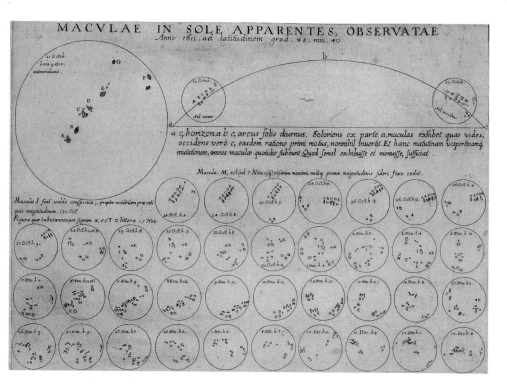

Above: This plate, originally published in the Rosa Ursina, depicting Scheiner's observations in Ingolstadt, appears in several different versions in early publications on sunspots.

Opposite: Various methods of observing the sun. Below: presumably Scheiner sitting while an assistant observes the projected image.

beautifully produced *Rosa Ursina,* is a polemical work undoubtedly directed against Galileo. In the book Scheiner defends himself by saying that Apelles never in word, deed, or suggestion hurt Galileo and at the same time violently attacks him. He questions Galileo's veracity concerning the dates of the observations. 'To talk about sunspots and not to write, to see sunspots and not to publish' is not worth anything, says he.[10] His tone is full of hate. When Castelli, admittedly Galileo's pupil and friend, saw the book, he was disgusted by the 'bestiality and poisonous rage' of the author. Scheiner made a list of 24 of Galileo's errors and arranged them in a column with the ironic heading: 'From the censor's (i.e. Galileo's) opinions.' In a second column he listed his own, as 'Apelles' propositions.' Galileo hated his mistakes aired in public. He wrote: 'This pig, this malicious ass, he catalogues my mistakes which are but the results of a single slip-up.'[11]

THE ENIGMA OF SUNSPOTS

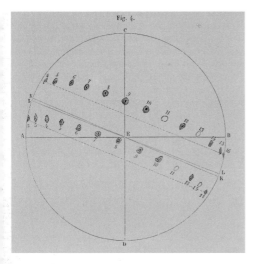

Fig. 4.

Scheiner spent sixteen years assiduously studying sunspots, sometimes making as many as twenty or thirty observations a day. He organized a network of observers, some living as far away as the West Indies, who sent him reports. Even so, giving the quality of his instruments as an excuse, he was unwilling (or unable?) to draw many firm conclusions. The 800 page long *Rosa Ursina* was four years in printing, and finally published in 1630. By then Scheiner accepted that the spots were on the Sun and that the Sun rotates. He set 27 days as the period of rotation. He established the orientation of

Spots in the "Royal Zone." the axis of rotation (about 7°). He confirmed that the spots occupy a definite zone around the Sun's equator north and south, can rarely be seen close to the equator, and cannot be seen near the poles. The zone where spots were most frequent Scheiner called the 'Royal Zone.'

Remember the Chinese astronomer whose report of an eclipse starting up in the middle of the Sun we assumed to be a sighting of a sunspot? Scheiner made the opposite suggestion: that the darkening of the sky, thought by some to be an eclipse at the death of Jesus was the appearance of an unusually dark spot. The Gospels of St Mark and St Luke both mention this

More spots appear within ±30° from the solar equator.

unusual event. In St Luke's words: 'And it was about the sixth hour, and there was a darkness over all the Earth, until the ninth hour. And the Sun was darkened.'

Two hundred years later the French astronomer Dominique François Jean Arago, whom we shall meet again later, remarked that Scheiner 'did not mean ... to deprive the phenomenon of its miraculous character. He merely thought of substituting an easy miracle for a difficult

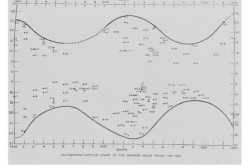

HELIOGRAPHIC LATITUDE CHART OF THE GREATER SOLAR SPOTS, 1881-1900

THE ENIGMA OF SUNSPOTS

one.'[12] In order to understand Arago's remark we have to keep in mind that total eclipses last for a few minutes and never for three hours as stated in the Bible, but spots can linger in the Sun for many days.

To give Scheiner his due, he was not the only one to suspect that large sunspots might be responsible for long-term dimness of the Sun. Two centuries on, Flaugergues was convinced that the dimness of the Sun which brought about an unusually cold year and crop failure in the year of Julius Caesar's death, as reported by Plutarch, was due to an unusually large sunspot. (It is now thought that it may have been the result of an eruption of the volcano Etna.)

In addition to his sunspot observations and calculations, in the *Rosa Ursina* Scheiner made a detailed study of the effects of various lens combinations, their advantages and disadvantages, and

also described his *helioscope*. This latter was a specially mounted telescope that enabled the observer to follow the movement of the Sun and projected the image onto a perpendicularly fixed piece of paper. A hundred drawings of sunspots (umbra and penumbra) and *faculae* accompany his text. But in spite of its lavish appearance the book was tedious, not a good read, and attracted bad reviews. The verdict of a nineteenth century historian was that in the 784 pages of the *Rosa Ursina* there was hardly enough material for fifty. Observations made in the past are more valued now than they were a hundred years ago but there is no doubt that Scheiner was running out of steam.

Not so Galileo. Although he had earlier stated that everything that could be said about sunspots had already been said (by him, of course), Galileo returned to the old topic once more, although

Scheiner's helioscope

not in order to make new observations. He used the spots and their behaviour in arguments both for the alterability of the heavens and for the Copernican system.

Before introducing 'these importunate spots' in the *Dialogue on the Two Chief World Systems*, the book that brought about his downfall, Galileo argues that the fact that in previous ages no changes were seen in the Moon and Sun is no proof of their unalterability. They are too far away for us to properly discern changes in them. Salviati, the character standing for Galileo, says sarcastically to Simplicio, upholder of the Aristotelian system: 'You must consider China and America celestial bodies, since you surely have never seen in them alterations which you see in Italy.'[13] Then Galileo reaffirms that many spots 'originate in the middle of the solar disc, and likewise many dissolve and vanish far from the edge of the Sun, a necessary argument that they must be generated and dissolved.'[14] A necessary argument but not a sufficient one. To dispose once and for all of the perfect unchanging Sun he had to show that the spots really are on the Sun's body and not somewhere in space around it. To do this Galileo uses another observation already mentioned in the letters and also used by Fabricius even earlier: 'Around the centre they are seen in their majesty and as they really are; but around the edge, because of the curvature of the spherical surface, they show themselves foreshortened.'[15] This agrees exactly with what would happen if the spots were located on the surface but cannot be explained if they orbit the Sun at a distance.

The argument relating to the Copernican system is somewhat more technical but the gist of it is easily understood. Galileo uses a philosophical expedient called Occam's (or Ockham's) razor. William of Ockham was a theologian in the early fourteenth century whose dictum is usually quoted as: 'Entities are not to be multiplied without necessity' although more accurately what he said was: 'What can be done with fewer [assumptions] is done in vain with more.' This principle is sometimes used to decide between two theories when both seem satisfactorily to explain the same phenomenon. Occam's razor and its application are often the subject of argument, especially between philosophers of science.

THE ENIGMA OF SUNSPOTS

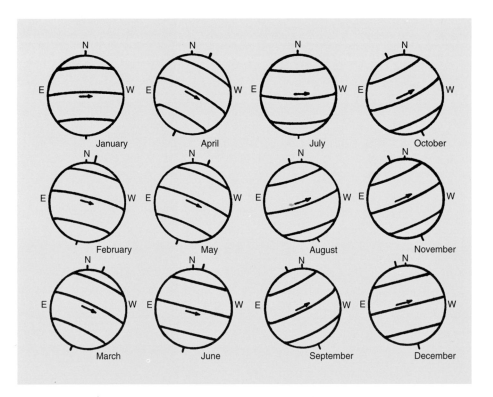

January   April   July   October

February   May   August   November

March   June   September   December

Relying on the observed paths of the sunspots Galileo shows that if we suppose the Earth being at rest, then the Sun would have to have four different motions, two to account for the movements of the spots and two for the annual and diurnal (daily) motion. We have seen that the way the spots moved was indication that the Sun revolved around itself (one movement). If its axis of revolution were to be perpendicular to the ecliptic then the spots 'would appear to us to [move] in straight lines, and parallel to the ecliptic.' This is not so, 'therefore the axis is tilted, since the courses for the most part appear to be made along curved lines.'[16] To top it all, those curves keep on changing over time and on an annual basis. This would mean that the axis of the Sun is not fixed but describes an annual path (second movement). Added to this would be the movements to

The movement of the spots follows the direction indicated.

account for night and day (third movement) and to account for the year (fourth movement).

Galileo's verdict was that:

> If these four motions so incongruous with each other and yet necessarily attributable to the single body of the sun, could be reduced to a single and very simple one, the sun being assigned one inalterable axis ... then really it seems to me that this decision could not be rejected.[17]

Of course, the four different motions by the Sun have not been reduced to one single motion because three are still necessary: one performed by the Sun revolving around itself, and two by the Earth orbiting around the Sun and revolving around itself. Some historians accused Galileo of sloppiness in the development of this argument and assessing Galileo in his book *The Sleepwalkers* Arthur Koestler found it 'obscure and incomprehensible.' Other historians think that Galileo was looking through some old notes when the book was nearly ready for the printer, found an old letter by an acquaintance telling him about the Sun's axis making an angle with the plane of the ecliptic and in his haste added the new proof without sufficiently developing it.

~ 7 ~

An Englishman Observes

In March 1584 Queen Elizabeth I granted Sir Walter Raleigh permission 'to discover search fynde out and viewe such remote heathen and barbarous landes Contries and territories not actually possessed of any Christian Prynce and inhabited by Christian people.'[1] Soon afterwards a reconnaissance expedition organized by Raleigh set out for America. When the members of this party returned at the end of the year they recommended Roanoke Island as a possible site for a colony. A second expedition, this time with a view of establishing a permanent base, left Plymouth in April 1585 under the captaincy of Sir Richard Grenville. Oxford graduate young Thomas Harriot was on board the main vessel, the *Tiger*. Harriot had been living in Raleigh's household as a tutor of navigation and mathematics. His duties during the expedition were to help with navigation, make scientific observations and survey the land. During his stay in America Harriot travelled more extensively than the other colonists, established friendly relations with the local Algonquian Indians and made an effort to learn their language. Wishing to establish the superiority of the white colonists and their religion, he suitably impressed the Indians with his scientific instruments.

Having spent a year on the island, and having managed to turn the originally friendly Indians into enemies by their insensitive and aggressive behaviour, the colonists were waiting for Grenville's return with badly needed supplies. Instead of Grenville, Sir Francis Drake turned up unexpectedly. Since he had heard rumours that the Spaniards were planning to root out the Virginian colony, he came in a hurry from Florida where he had just destroyed a fort. As he was approaching the coast, a hurricane blew up, damaged his fleet and prevented him from landing.

Drake's supplies were not sufficient to cover both the needs of his crew and that of the colonists for another year. In any case, the hurricane had by then convinced the latter that they should take advantage of Drake's offer of a passage home and when the wind abated they left with him in a hurry. When Grenville arrived somewhat later, he found to his surprise and dismay, that the site was abandoned.

On his return Harriot published the extremely successful *Briefe and true report of the new found land of Virginia*. It was the first English book exclusively about America. In this joint venture with John White, the expedition's artist who illustrated the 1590 edition of the book, Harriot describes the plant, animal and mineral resources of the region together with the customs of the Indians he encountered. Among other commodities he praises tobacco because: 'It purgeth superfluous fleame and grosse humours, and openeth all the pores and passages of the body ... [the natives'] bodies are notable preserved in health, and not know many grievous diseases, wherewithall we in England are often times afflicted.'[2] Sadly, cancer of the nose, a 'grievous disease' that carried off Harriot, was probably the result of tobacco smoking.

Back in England, Harriot found himself a new patron in the Earl of Northumberland. He moved to a house next to the main building of Syon House, Isleworth, with a generous salary of £300 per annum. Here Harriot occupied himself with scientific pursuits. He set up an observatory and by August 1609 already owned a telescope. In time his instrument maker, following Harriot's directions, constructed for him several telescopes of varying power. It seems that for his solar observations Harriot did not use the camera obscura, only the telescope when the sky was cloudy or misty. Consequently but not surprisingly he complained about problems with his eyesight. He did use coloured glasses for a while but later abandoned this practice probably because the glass was of inferior quality and distorted the image. His first notes and drawings of a sunspot date from December 1610.

The Earl was implicated in the Gunpowder Plot and was sent to the Tower for fifteen years. Harriot was also briefly imprisoned; he

THE ENIGMA OF SUNSPOTS

was kept in the Gatehouse. Life in the Tower was not unduly harsh for the prisoners. The Earl 'kept a handsome table there for Harriot and his mathematical friends' who were allowed to visit him.

Harriot never published the result of his investigations and one of the reasons for this may be found in the political situation of the time. Already at Raleigh's trial the Chief Justice described Harriot as the devil who brought Raleigh to atheism, and at the trial of the Duke of Northumberland the Attorney General took exception to Harriot's drawing up a horoscope for the King. In his will however, Harriot instructed his literary executor to publish his papers. His wishes were not complied with and only a part of his mathematical papers was published and the rest, amounting to thousands of sheets on astronomy, navigation, optics, and cartography, did not see the light of day.

At Harriot's death his papers became the property of the Earl of Northumberland. When that male line became extinct, one female descendant married into the house of the Earl of Egremont and the papers were transferred to Petworth House in Sussex. These papers stayed at Petworth. They were discovered only a hundred and fifty years later, hidden underneath a pile of stable accounts, by the Hungarian born Baron Franz Xaver Zach.

Described as arrogant and of boundless ambition, there is no doubt that Zach was of a gregarious and flamboyant personality. He was also a man of intrigue who slanderously accused several of his 'friends' of falsifying their observational data.[3] What brought him to Petworth? At this time, Zach had a position as tutor in the family of the ambassador of Saxony to London and during 1784 he spent a holiday with his young charge at Petworth, where he came across Harriot's manuscripts. Realizing the importance of his find, he hurriedly selected and took some of the papers. He then 'began to build his own reputation by making sensational claims' about the contents of the papers.[4]

Zach submitted a proposal to the University of Oxford for publishing a biography of Harriot, together with the astronomical papers in his possession. The proposal remained just that: a proposal. Zach published only a few notes but nothing of real value. But Zach's neglectful treatment of the Harriot papers is not to be ascribed to general laziness or ignorance, rather to the fact that his priorities were elsewhere. From 1786 onward he was in the service of Duke Ernst of Saxe-Coburg and in charge of the observatory the Duke had built for him near Gotha. He founded and edited a monthly journal in which he published the latest results in astronomical, geographical, hydrographical and statistical research. He organized the first international cooperation between astronomers, and has a solid list of scientific publications to his name. These activities were obviously more important for him than wading through the two-hundred-year-old papers of an unknown Englishman.

Ten years after he first set eyes on them, Zach sent the papers in a dreadful mess to Oxford. University staff started to organize and edit the manuscripts but the task was never finished. When Oxford University was criticized for failing to publish Harriot's astronomical papers, two professors defended the University by expressing the opinion that publishing the papers 'would not contribute to the advancement of science.' Abraham Robertson and Stephen Peter Rigaud, although impressed by the quality of Harriot's work, 'felt it necessary to disparage the originality of Harriot's contributions in order to justify the Oxford failure to publish the papers and to discredit the exaggerated claims of Baron von Zach.'[5]

Oxford University returned the papers they had received from Zach to the Earl of Egremont. He in turn gave the rest of the manuscripts that were still in his possession to the British Museum. The papers that came back from Zach via Oxford stayed at Petworth, presumably because the Earl thought that they were the more valuable ones. The Museum had the papers, although they were also in complete disorder, bound into eight tight volumes. The ones at Petworth were seen a few times more, in particular in the 1850s when the Swiss astronomer Rudolf Wolf of whom we shall speak later, needed historical data in support of a theory. Afterwards they lay more or less forgotten again for another hundred years until in 1948 at the insistence of an American scholar they were copied and made available. Since

Harriot's first drawing

THE ENIGMA OF SUNSPOTS

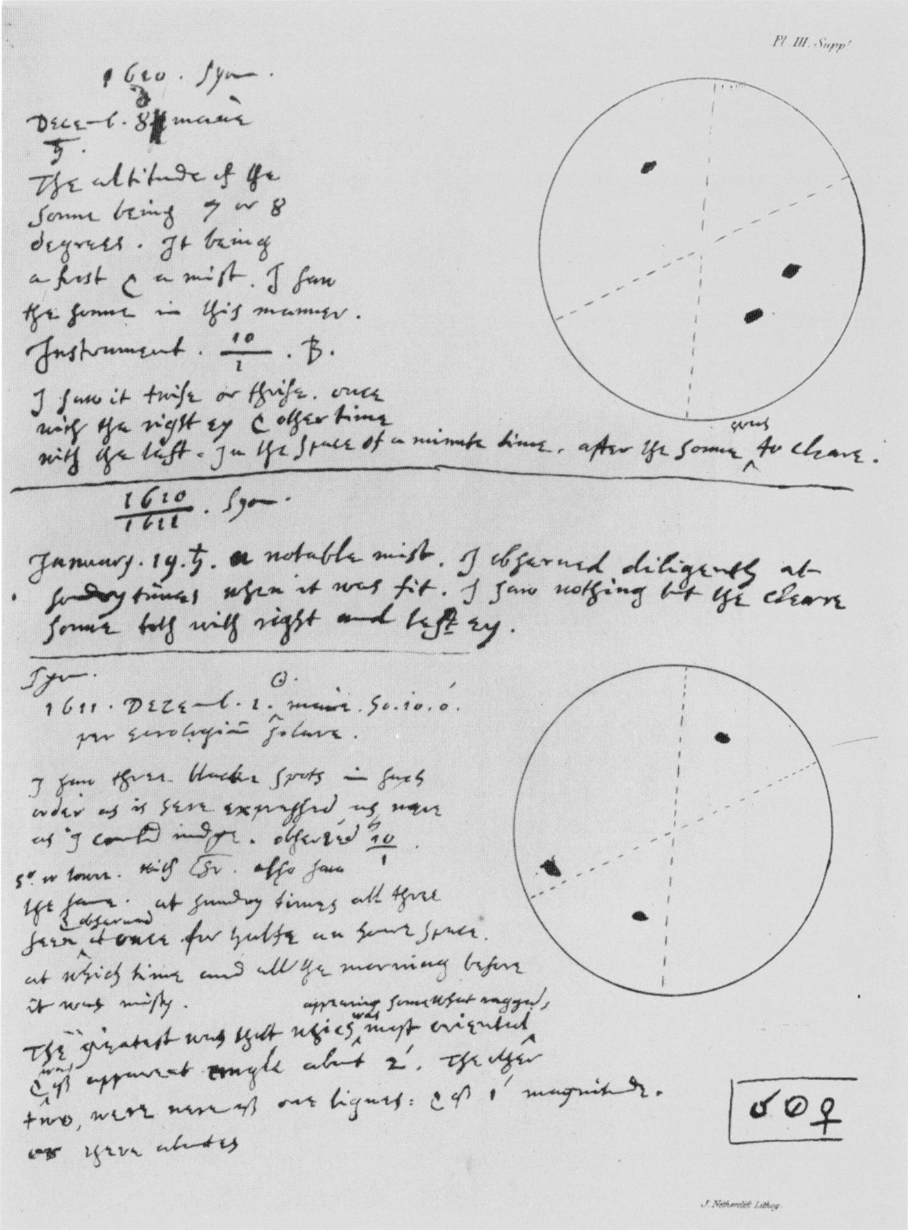
Pl. III. Supp!
1610. Syon.
Decemb. 8 mane ♄.
The altitude of the sonne being 7 or 8 degrees. It being a frost & a mist. I saw the sonne in this manner.
Instrument. 10/1. B.
I saw it twise or thrise. once with the right ey & other time with the left. In the space of a minute time. after the sonne was to cleare.
1610/1611. Syon.
January. 19. ♄. a notable mist. I observed diligently at sundry times when it was fit. I saw nothing but the cleare sonne both with right and left ey.
Syon. ☉.
1611. Decemb. 1. mane. 50.10.0. per horologium solare.
I saw three blacke spots in such order as is here expressed as nere as I could iudge. observed 10/1.
at sundry times all three I observed seen at once for halfe an houre space. at which time and all the morning before it was misty.
The greatest was that which was most orientall appearing somewhat ragged, & was of apparent angle about 2'. The other two, were nere of one bignes: & of 1' magnitude. or there abouts
♂ ☉ ♀
J. Netherclift Lithog.

One of Harriot's drawings.

then interest in Harriot spawned a whole little research industry. Although their enthusiasm seems to have abated by now, in the 1960s and 1970s the 'Harrioteers' devoted themselves to researching Harriot's life and work.

Some of his eulogists regard Harriot as the first discoverer of sunspots since — even making allowances for the calendar difference between England and the Continent — his first note about them dates from December 1610, earlier than either those of Fabricius or Scheiner. But though it is true that Harriot first drew sunspots then, he started regular observations only a year later, after the publication of Fabricius' little volume.

Whether or not we accept Rigaud's criticism that Harriot's work did not reach the highest calibre, and whether or not we accept that he discovered the spots, Harriot's work on the Sun is still important. His notes contain records of about 450 observations and about 250 drawings of sunspots and they give indication of how much the Sun's surface was covered at the time, how the spots changed and how long they lasted. He always took his instruments with him when travelling to London (Isleworth where he lived, is now a suburb but then it was a settlement far removed from London) or anywhere else, so we can assume that he observed quite regularly. Nevertheless, Harriot's notes do present problems. Only fair copies of his drawings have survived. Often when there are certain days missing from his observations we can only guess why this is so. Was he ill, was the day cloudy or, was he not satisfied with the original drawing or, were there no spots to be seen? Some guesswork had to be incorporated into the historical data, but even so, Harriot's observations were useful in establishing the sunspot cycle and in our time they can be used to build theories not only about the structure of the Sun but also about its terrestrial influence.

THE ENIGMA OF SUNSPOTS

~ 8 ~

Sunspot Watching as a Popular Pastime

Here we might pause for a minute to consider what exactly is a discovery and what counts as a discovery for the scientific community. Galileo turned his telescope towards the night sky and saw many more stars than he could see with his naked eye. Undeniably he discovered new stars hitherto unseen and unknown. In 1612 his German contemporary Simon Marius (or Mayer) turned his telescope towards the night sky and saw a distant nebula (a cluster of stars outside our galaxy). This nebula, the Andromeda, can be discerned with the naked eye and most probably people with sharp eyesight had seen it previously, but since they never reported it, Marius has been credited with its discovery.

On the other hand, sunspots were not only seen previously, but the sightings had been documented, fairly extensively in the East, less in Europe. The question arises: when the astronomers in the seventeenth century turned their telescopes towards the Sun and noted its spots, did they really discover something new? Does not the honour of discovery, which was so bitterly disputed, belong to a Babylonian astronomer by right?

We could argue that observation in itself is not sufficient. The discoverer has to disseminate the news of the discovery to a wide audience. Arago was of this opinion. He thought that a discovery is not a discovery if it is not published in one form or another. He explained:

> I also call publication all academic lectures, in fact all lectures given to a wide public, all printed reproductions of the discovery. But private communications do not have

the necessary authenticity. Certificates given by friends are worthless.[1]

The last sentence refers to Galileo, without a doubt.

Does a sighting count as discovery if no explanation is forthcoming? The logbook of the good ship *Richard of Arundell* recorded a sunspot and later this was duly published by Hakluyt; as a result it satisfied Arago's criteria. But the sailors were busy and had no time or inclination, and probably not enough education, to discourse about patches on the Sun. They did not realize the possible significance of what they saw, a common enough occurrence in everyday life. We should not therefore reproach these early observers, but we cannot regard them as the true discoverers. On the same grounds we can disqualify Harriot out of hand as well, because not only did he not publish anything but he did not even try to give an interpretation to his observations. At least not in writing — if he had any thoughts on the matter, he kept them to himself.

Regular observation (or experimentation if that is possible) is also an important factor. If a scientific discovery can not be reproduced, it will not be accepted. Similarly, meaningful conclusions can not be based on a single astronomical observation.

Who then did discover sunspots? Johann Fabricius was the first to make his findings public by writing and publishing a book in 1611. He made regular observations over a period of time and tried to explain, or at least to interpret what he saw. Our criteria are satisfied. Galileo may have seen the spots earlier and shown them to his friends, but obviously he did not find them worthy of further attention prior to reading Scheiner's letters. The timing of Scheiner's first observation is not well defined and he in his turn also neglected the blemishes right until after Fabricius' book was published.

Priority of discovery may not be an important question if we consider only the progress of knowledge, but from a personal, financial and national aspect it is extremely important. Galileo was especially jealous of his intellectual reputation — and in 1607 even took one of Marius' pupils to court for plagiarism —

but who would not feel a glow of satisfaction from being the first to discover something? Financial rewards in Fabricius' time meant patronage that could depend on personal prestige, while in our time patents (the possibility of acquiring money), further research grants (money) and prizes (prestige and sometimes money), all depend on priority. National pride cannot be forgotten either. English sources often hold Harriot the discoverer of sunspots.

In the first half of the seventeenth century the Sun and its spots were often watched by many curious souls. The invention of the telescope and the new discoveries made with it created immense interest in sky gazing. The very public arguments, conducted not only between Scheiner and Galileo but also between their camp followers, led to the Sun and its spots being observed with extra care. But after the first flurry of discoveries, with few exceptions no new facts regarding the Sun emerged, and no significant new theories about its spots were forwarded, rather the old ones were chewed over. It was time for consolidation.

Pierre Gassendi.

One of the more enthusiastic observers was Pierre Gassendi, priest, mathematician, natural philosopher. He listed the results of his daily industrious observations of sunspots and reaffirmed most of the by then known facts. Gassendi saw the spots in a band close to the equator of the Sun but could not find any at the poles, he saw some being born on the disc of the Sun and some perish before they left the disc. The spots he saw were irregularly shaped but invariably fatter and rounder when they were in the middle of the Sun than at its edge (the limb). All were moving faster in the middle and slower at the limb. He was not the only one to believe, wrongly, that spots moved with equal speed in all latitudes. One of Gassendi's claims to fame is that he was the first person who saw a transit of a planet. This required extraordinary

Nicolas Claude Fabri de Peiresc.

Gassendi's observations.

tenacity. Kepler predicted a transit of Mercury for November 7, 1631, but was not sure that his calculations were correct. He therefore recommended that astronomers should begin their observations a full three days before the predicted date. Gassendi and his good friend Nicolas Claude Fabri de Peiresc excitedly prepared for the event. Peiresc, an enthusiastic amateur astronomer who had installed an observatory on top of his house, lived in Aix-en-Provence in the sunny South of France; Gassendi was in cloudy Paris. It was likely that Peiresc would be able to see the transit but Gassendi would not. It turned out the other way round.

Gassendi projected an image of the Sun in a dark room (*camera obscura*) through a telescope. This was the same method he used for his sunspot observations. Following Kepler's advice he sat down by his instrument and waited. On the first day it rained. Mercury may have passed unnoticed but Gassendi did not give up. The second day was foggy and cloudy. Again, Mercury may have passed unnoticed but Gassendi still waited. On the third day there were some sunny periods. Just when the sky cleared, luck at last: a small spot appeared in front of the Sun. By then Gassendi despaired of seeing Mercury and was convinced that the fleck he saw was an ordinary sunspot. He expected Mercury to be much larger than that! But the spot moved too fast and in the end Gassendi changed his mind and decided that it must have been Mercury after all. He quickly measured the diameter of the round spot and stamped his feet. By agreement an assistant in the room below should have heard the stamping and should have quickly measured the height of the Sun above the horizon so that the exact time of the observation could be recorded. But the young man had got fed up with all that waiting and gone home by then. Gassendi had to be satisfied with his measurement of Mercury's size. Kepler

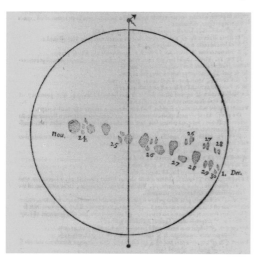

THE ENIGMA OF SUNSPOTS

mistakenly thought a spot was Mercury; Gassendi mistakenly thought Mercury was a sunspot; but from now on the size and the shape of the planet was known and there was an end to confusion.

While Gassendi was at his vigil, Peiresc on the other hand ... In Gassendi's words, he had set up:

> [a] scene on to which the rays of the Sun were to be let through a prospective glasse that the appearance of the Sun being exhibited, the shadow of Mercury might be discovered. And indeed he watched as much as he could, or did remember ... But the apparition ceasing before noon, the presence of certain guests made him forget it all morning, for he accompanied them to hear Masse and feasted them afterward liberally. He complained therefore exceedingly when he was acquainted with the time in which the spectacle appeared.[2]

Well, you either watch the sky or entertain your guests. But Peiresc, a former student of Galileo, a wealthy man, was a member of the Provençal parliament, who had onerous social and official duties. He was so influential that in order to speed up his correspondence, at his request the Provençal government hired special riders who could make the return trip to Paris in two weeks. Peiresc was also a collector of books, manuscripts, coins, medals, curiosities of the natural world, astronomical instruments and art objects. In fact he collected anything that caught his fancy. He is sometimes labelled a 'connoisseur' and there can be no doubt that Peiresc was that in the modern English sense of the word as a well informed judge and lover of beautiful things.

Peiresc was also a patron of the sciences and his house served as a meeting place for scientists and artists. He was one of the hubs of the international scientific community. It was thanks to Peiresc that much of contemporary Italian science filtered through to France. But astronomy was only one of Peiresc's many interests and it is no wonder that he was not as regular an observer of the skies as for instance Gassendi was. In fact, his working methods strike us as distinctly casual.

Opposite above:
*Mrs Hevelius was
an able assistant to
her husband.*

Opposite below:
*Hevelius draws the
projected image of
the sun during an
eclipse.*

Below:
*Hevelius' rooftop
observatory.*

By contrast, the grand amateur, Johannes Hevelius, devoted his entire life to regular sky watching. Northern skies are not conducive to stargazing but the inconvenience of freezing cold nights and the frustration caused by frequent cloud cover can never deter the resolute as the example of Tycho shows. Hevelius modelled his life on Tycho. Born in 1611 in Danzig, the once disputed territory between Germany and Poland (it belongs to Poland now, as Gdansk) Hevelius was fluent in both German and Polish. The family was rich and he was sent to study law with the expectation that he would become an important city official.

His parents owned a brewery which he later inherited (there seems to be an affinity between brewers and astronomy). As a magistrate and city councillor, Hevelius at first tried to juggle his official and business duties with his scientific interests but eventually managed to make astronomy more or less his full-time occupation.

Hevelius started out by building a small observatory in a room on the top of one of his many houses. Then he added a small tower. Then he added a 15,000 square foot platform on which he built two huts, one of them revolving. Eventually the whole contraption extended to three rooftops and figured as a famous tourist attraction. Hevelius ground and polished his own lenses for some of his telescopes and being a good engraver he engraved the scales of his brass instruments with great precision. He was also an accomplished draughtsman and many illustrations in his books are his handiwork.

Hevelius worked round the clock: he watched the starry sky at night and made his calculations, his drawings and engravings by day. His second wife was one of his best assistants and in due recognition of her services to astronomy she is shown helping her husband in a couple of pictures in one of Hevelius' books.

Hevelius is rightly famous for being the first astronomer to produce a complete and accurate map of the Moon. He

EDMUNDUS HALLEIUS R.S.S.
Astronomus Regius et Geometriæ Professor Savilianus.

Edmund Halley.

also set himself the task of compiling a catalogue with the most accurate tables of the positions of the stars. He prided himself on his good eyesight and claimed that naked eye observations were better than those made with a telescope. Instead of fixing small telescopes onto his quadrant and his sextant, he used simple rifle sights. This resulted in a flaming row with the Royal Society in London and especially with Robert Hooke. In his book bearing the wonderful title: *Animadversions on the first Part of the Machina Coelestis of the honourable, learned and deservedly famous astronomer Johannes Hevelius,* published in 1674, Hooke says that 'So great a curiosity as Hevelius strives for is needless without the use of telescopic sights,' and if Hevelius 'could have been prevail'd on ... to use telescopic sights, his observations might have been forty times more exact then [*sic*] they are.'[3] He doubted that they were any better than Tycho's. Hevelius called Hooke's bluff by asking him to make and send him such observations. Then young Edmund Halley, recently back from St Helena where he was charting the southern skies, was sent to Danzig to smooth matters over and to examine the instruments Hevelius used and his results. Halley stayed for two months and the two astronomers worked together. They made measurements with telescopic sights and without. Mrs Hevelius helped too and, if we can believe the gossiping tongues, she struck up more than a Platonic friendship with Halley. The outcome of the visit was inconclusive. It showed that Hevelius had an excellent eye and that the instruments of the time still needed improving.

Telescopes were still optically quite deficient at the time, the lenses distorted the image in more ways than one, blurred the

THE ENIGMA OF SUNSPOTS

outlines and produced coloured fringes. The trouble was that the large telescopes Hevelius built were even worse than average. Some were so long that their wooden tubes warped from the humid air, they shook in the slightest breeze, and the lenses kept coming out of alignment. No wonder he trusted his own eyes more.

For his sunspot observations Hevelius used more satisfactory small telescopes and a camera obscura. He also used a scioptric ball. This was a 'lens with a swivel mount' invented in 1636. Often fitted with a mirror to project the image, it could be swivelled around to follow the movement of the Sun.*

Progression of spots as observed by Hevelius

It was rather unfortunate that Hevelius did not publish a separate book on his sunspot observations, but tucked them — together with some wonderful engravings he made — into his book on the Moon and into his other book on the comets. This meant that his work was often overlooked when people were looking for historical data. Then, when it was realized that in Hevelius' time special circumstances prevailed, suddenly his observations became important. We shall see later what these special circumstances were.

A few months after Halley's visit in 1679, fire destroyed practically all of Hevelius' possessions. His seven houses, his observatory, the printing press and the library, all went up in flames. Although Hevelius built a new observatory and equipped it with new instruments, his work soon came to an end.

* Sometimes referred to as the sky optic ball.

～ 9 ～

Facts and Speculations

In the middle of the seventeenth century there were still a few dissenting voices. Some, who could not accept the imperfection of heavenly bodies claimed divine intervention. Their argument was *Spots orbit around* that for reasons we are unable to fathom, the Almighty created the *the Sun (on the two* occasional blemish. Some others still stuck to the belief that *innermost circles)* sunspots were small planets circling the Sun. Among them was Otto *in this diagram* von Guericke who is better known for constructing an air pump *from* Experimenta and proving the existence of a vacuum with his 'Magdeburg Hemi-Nova *1672, by Otto* spheres.' The generally accepted version however, was that spots *von Guericke* were actual material features on the very surface of the Sun.

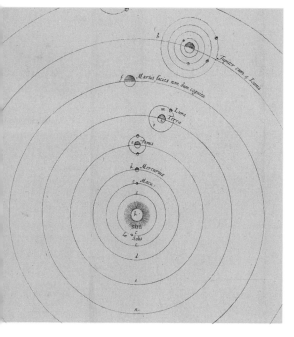

By the time the transit of Venus was first observed in 1639, the argument had already turned around. Previously the emphasis was on proving that the spots were not planets, now it had to be shown that a planet was not a spot. The first to see Venus pass in front of the Sun and identify it as such was Jeremiah Horrocks (or Horrox), a young amateur astronomer. Horrocks died very young and most of his notes went up in flames but luckily we have his description of the transit. It took place on a Sunday when Horrocks as a clergyman was busy for much of the day so he was lucky to be able to see it. Horrocks was also a

78

THE ENIGMA OF SUNSPOTS

sunspot watcher and knew the difference between their appearance and behaviour and that of a planet. He was certain that he saw Venus transiting the Sun because it had a regular circular shape and did not move any slower at the edge but crossed the disc with constant speed. Horrocks was not confused as Gassendi had been, he knew that what he saw was a planet and not a spot.

By this time the most important characteristics of the spots had been described in detail. Most spots consist of a dark nucleus and a relatively lighter penumbra. They grow faster than they dissipate, and often appear in pairs or in larger groups while others stand single. Some die young after being born and never have time to make a full revolution. Others persist and after disappearing at the western edge of the Sun, reappear on the eastern side. Some split into two, some others merge together. Spots are fatter and rounder in the middle of the disc than when they are at the edge. They assume different sizes: in Galileo's lively description some 'are so vast as to exceed not only the Mediterranean Sea but all of Africa, with Asia thrown in.'[1] The really large ones can cover billions of square miles and be seen with the naked eye. We see spots mostly in a band close to the solar equator.*

In the years following their discovery regular observations, drawings and measurements yielded a great deal of new and useful information, not only about the spots but also about the Sun. The spots' movement from east to west on the Sun's disc proved that the Sun rotates. By following their progress the angle that the axis of rotation makes with the ecliptic could be calculated. The spots did not all move with equal speed, those closer to the equator moved faster than those further removed from it, indicating that the outer layer of the Sun cannot be solid. This made the calculation of the speed of the rotation difficult and imprecise. Those who based their calculation on spots nearer the equator found that it took the Sun about 25 days to revolve around itself, while those who measured the speed of revolution using spots at high latitudes put it at more like 29 days.

All this was known already within the first fifty years of the orig-

* Scheiner's Royal Zone, see p. 58.

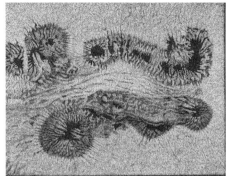

Secchi's drawings of sunspots.

inal discovery, and still nothing could be said about the chemical composition or the physical nature of the Sun and its spots for another two hundred years. Even an earlier, reasonable suggestion by one of Galileo's pupils, that the spots may be cooler than their surrounding region could not be verified because there was no method of measuring their temperature.

In the long interval between the discovery of sunspots and the first glimpses of the physical and chemical nature of the stars, i.e. the birth of astrophysics, inevitably speculation flourished. Marius summarized the chaotic situation that existed during his time, soon after the original discoveries:

> It had been my intention, according to my former proposal to deal now with the spots on the Sun, setting out my observations upon them from August 3, 1611, to the present time. However, I do not wish — and indeed, am unable — to make any definite statement about them at present, ... for the ... reason that I find the greatest authorities in disagreement and am unable to satisfy myself.[2]

In 1621 Robert Burton expressed a deeper general dissatisfaction with the state of affairs in his *The Anatomy of Melancholy:*

> Our later Mathematitians have rolled all the stones that may be stirred, and to salve all appearances & objections, haue invented new hypotheses ... one saith the Sun stands, another he moues, a third comes in, taking the[m] all at a rebound: and lest there should any Paradox bee wanting, he findes certain spots or clouds in the Sun, by the help of glasses, by means of which the Sun must turne round upon his own center, or they about the Sun. Fabricius puts only three,& those in the Sun, Apelles 15 & those without the Sun.[3]

One thing was certain: the outer layer of the Sun could not be solid because otherwise the spots could not be seen changing shapes. Everything else was up for grabs. Some held that the Sun was a fiery ball with an atmosphere and its spots were like terrestrial clouds: this was of course Galileo's original view. The outer layer of the Sun was like a boiling liquid and the spots were the scum on it, said others. No, they were like slag on molten metal. Then again, the interior of the Sun was cool, much like our Earth and the spots were smoke and fumes: results of volcanic eruptions from deep inside. For those in the opposite camp they were high mountains normally hidden by opaque layers and occasionally suddenly uncovered.

The origin of comets had always been the subject of speculation and they came in handy for new theories that would account both for comets and spots. If, as it was believed, comets formed from material that escaped from the Sun, or from another star, then a simple explanation presented itself: spots were the very material comets were made of. At first the stuff assembles on the Sun, or on a star and appears to us as a sunspot or a starspot. (We do not have direct observations of spots on other stars. Regular variation in brightness would indicate that some have spots, but this is only an assumption.) Eventually the material is expelled and becomes a comet. When in the relatively spotless year of 1618 three comets appeared, this particular theory seemed to be confirmed. Nevertheless, not everyone agreed. The contrary opinion, that spots were comets, or small planetary material arriving from space, falling into the Sun and gradually dissolving in it, was also attractive and acceptable.

There may not have been agreement about the nature of sunspots, but all the analogies used in the speculations about them were based on terrestrial experience. There was no suggestion that different material from that found on Earth would be responsible for the spots. The discovery of the spots showed that the Sun is not a perfect body and their behaviour showed that the stuff the Sun is made of is subject to alteration. Galileo's views were influential, but even without him further argument against Aristotelian dogma turned out to be wholly unnecessary. As if by

René Descartes.

stealth, due to the discovery of the spots, the Sun ceased to be a heavenly body made from a completely different material than the Earth and suddenly it was subject to the same physical laws that govern terrestrial material.

We shall see later that this belief — because belief it is — took some knocks three hundred years later. In the seventeenth century however, the existence of sunspots led to the acknowledgement of the unity of the heavens and of Earth. This is nowhere more apparent than in René Descartes' curious theory about matter in the universe. Descartes, a French philosopher, formulated two distinct versions of his theory and in doing so he added to the already burgeoning body of speculations. In the first version heaven and Earth were made of different materials, while in the second version heavenly and earthly material could interchange. I think that his change of mind was due to the increasing importance he felt he had to assign to the sunspots.

According to Descartes' second version, printed in his *Principles of Philosophy* in 1644, the world in its present form is made up of three elements, but was not created as such. The particles of the single original element were equal in size but irregularly shaped. Because in Descartes' world everything is constantly moving round and round in a vortex motion, particles keep on rubbing against each other and become rounded. Descartes called these little balls 'element number two' in the present state of the world. The bits that were rubbed off the original particles are very small, irregular and agile. This débris fills the space between the now rounded particles of element number two, and is 'element number one.'

It can happen that more number one element particles are made than needed to fill the space between the round globules of

THE ENIGMA OF SUNSPOTS

element number two. (The reader may wonder how this is possible!) These congregate in the middle of the whirlpool and become stars or, the Sun in our solar system. Now, because of the constant circulation of matter, the stars as it were 'bubble up' at the poles and the material moves towards the equator. As it flows, larger particles stick to each other, and the whole thing floats like scum on a boiling liquid. Forming very large dark masses, element number three is thus created. This is what sunspots are.

Descartes based his theory on the copious information about sunspots he received from his correspondents. Outlandish as it may seem, this theory turned out to be quite useful in explaining their properties. The spots are seen in a band fairly close to the equator because although the material bubbles from the poles, it takes some time for it to congeal into a large enough mass to be seen. The spots vary in size and shape due to the haphazard way they stick together. Some are destroyed 'In the same way as many liquids, by boiling longer, reabsorb and consume the same scum which they gave off at the beginning by bubbling up.'[4] Anyone who has ever made soup and was too lazy to skim it while cooking, can attest that by and large this is a true description of what happens. Faculae are places where the first element is in a tight space and surges with great speed.

Using his theory of matter Descartes uses his fertile imagination to explain not only sunspots but variable stars, planets and comets as well. When first formed, spots are soft but if they survive for a while, then their inner surface, the one next to the Sun's body, becomes hard and polished and the Sun (or star if they are starspots) is unable to reabsorb them. Such elderly spots can eventually cover the whole surface of a star, preventing it from emitting any light. This star then disappears from our view. It can reappear again, because the spots are porous and let the material from the centre re-emerge through the pores. The star reappears when enough material has escaped, and its surface is covered once again with the first element. This process repeats itself and the star ends up like an onion with an inner core of first element substance, and then alternate layers of third element spots and first element material. Occasionally, when a star collapses, it can be completely engulfed by third element material. Then the star becomes a comet or a planet. In Descartes' theory there is no basic difference between the material the heavens and the planets are made of; in fact they constantly interchange. The Sun is one of the stars and the stars are other suns. This final conclusion is the view we hold today although we do not subscribe to the rest of Descartes' speculation.

So we may agree with the Dutch scientist, Christiaan Huygens, who wondered how Descartes, 'an ingenious man could spend all that pains in making such fancies hang together.'[5] But it would be a mistake to look on all the speculations about the Sun and its spots as fancies of deranged minds. Science would never progress without this kind of surmise, without speculation and imagination. As long as the conjectures explain some of the observed phenomena and do not obviously contradict any of the others, speculation is permitted, even encouraged. It could be the spark that ignites.

Old products of imagination sometimes find an echo in modern science. This is how two hundred years after Descartes, the astronomer Rudolf Wolf was trying to explain a phenomenon related to him by the British astronomer Richard Carrington: 'I compare the whole appearance of sunspots to currents which flow periodically from the two poles of the Sun towards its equator.'[6] Was he cribbing from Descartes without acknowledgement? I don't think so, I rather think that he came to a similar conclusion, but via a different route. Old ideas are constantly recycled, newly interpreted, given new functions, new explanations and sometimes they are given new names. We still think that the Sun has an atmosphere, it is just that we do not think that it is like our atmosphere here on Earth. On the other hand, the old term 'exhalation' has been renamed 'solar wind.'

Earlier speculations of course, extended to the terrestrial influence sunspots may have. Meteorological connections from earlier ages have already been mentioned. The Jesuit Athanasius Kircher attributed a more profound significance to the spots than rainmaking facility. Kircher, ordained in 1628, was at first professor of philosophy, mathematics, Hebrew and Syriac (a language used by Christian writers in the Middle East), in Würzburg. During the Thirty Years War he escaped from Germany, spent some time in Avignon and finally ended up teaching in Rome. Not only was Kircher interested in everything, he knew everything about everything as well. A real Renaissance man, he wrote volumes on physical geography, on the plague, on magnetism, and on music to name but a few of his topics. He attempted to decipher the Egyptian hieroglyphs and to construct a universal language. His curiosity was such that he had himself lowered into the crater of the volcano Etna. The Museum Kircherianum in Rome held his massive collection of instruments, curios and antiquities until the end of the nineteenth century. The contents were then redistributed among the other museums in the city.

Athanasius Kircher.

Kircher was held in great esteem for his erudition. Hevelius visited him in Avignon and afterwards their letters are full of reciprocal admiration. He was also a friend of Peiresc who invited him to Aix, where he met Gassendi and where they held long discussions on the nature of sunspots. Kircher served as another hub in the seventeenth-century network of scientists; at the latest count he was found to have had over 760 correspondents. They sent him information and contributions to his collection. Kircher reciprocated with balsams and other precious medical remedies.

In April 1625 Kircher looked at sunspots through a telescope and was so overwhelmed by what he saw that 'from that day onwards astronomy became one of his chief studies.' Ten years later, in Rome he studied the Sun jointly with Scheiner. Kircher did not understand how Aristotle could envisage the Sun as a cold body. Looking through Scheiner's helioscope it was obvious

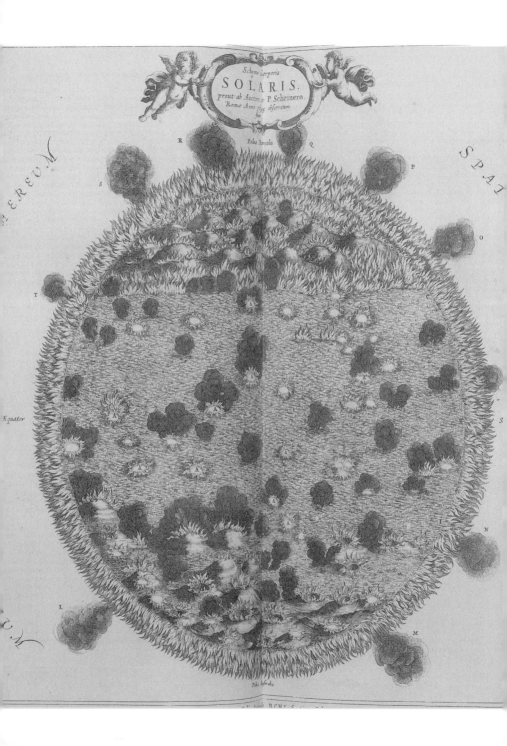

to him that the Sun was like a fiery furnace, a billowing sea of flame. He reasoned that if something appears to be like fire and has the same effect fire has, well then, it must be fire. The spots are vast evaporations from this ocean of fire and he could see them forming and disappearing. The question remains: could Kircher really see flames on the Sun's disc and spots appearing from the flames? Did he report rather what he wished to see, instead of what he actually saw?

Kircher's emphasis was on the influence of sunspots on Earth. He urged the astrologers to study the sunspots instead of the stars because he believed that anything they might say about the constellations should in reality apply to sunspots. Eruption of a great number of spots was for him a forewarning of terrible things about to happen. As a good example of this he cited the appearance of spots preceding a Swedish invasion of Germany.

We must not ridicule Kircher. First of all, the Sun does influence us directly and there is no intrinsic reason to suppose that sunspots do not do so. Secondly, changes in the Sun might have a physical effect on us and sunspots are sure signs that the Sun is undergoing some changes.

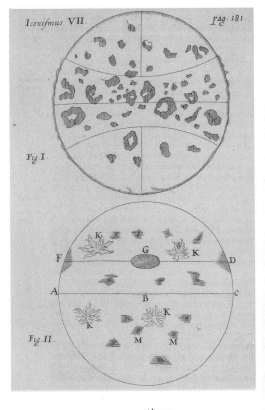

Above:
Sunspots and faculae by Kircher.

Opposite:
The surface of the sun according to Kircher.

A Breakthrough

At first the many contradictory hypotheses encouraged further research and the situation seemed hopeful: information collected about the sunspots would eventually lead to more knowledge about the Sun itself. In 1700, the French astronomer De La Hire bluntly stated that if ever anything will be discovered about the nature of the Sun, it will be through its spots. He was wrong. Although the observation and study of sunspots is still vitally important, it is only one of several methods available to us in our search for a deeper understanding of the Sun.

Strangely enough, De La Hire made his announcement when interest in sunspots was at a low ebb. Was their study only a short-lived fad? Or were there good reasons for astronomers to turn to other topics? Spots were regarded by many as mere curiosities and this could be one reason for the relatively few sunspot observations in this period. Science thrives on regularity, a pattern always hints at connections to be discovered and demands further study and explanation. Sunspots stubbornly refused to exhibit any regularity. In 1613 Johannes Kepler had this to say about the freckles:

> Like clouds they appear, divide, dissipate, disappear.
> Likewise we see some tiny spots rise in the middle of
> the Sun's disc, in successive days they will grow, others
> rarefy and vanish before reaching the extreme margin,
> some divide into two or even three parts.[1]

Or, as Agnes Clerke wrote in her scholarly and influential *Popular History of Astronomy during the Nineteenth Century*: 'The reckoning and registering of sun-spots was a task hardly more inviting to an astronomer than the reckoning and registering of summer clouds.[22] There were of course, those like Kircher, who believed that sunspots might have an effect on human life. At times the appearances of spots (and of auroras too) were supposed to be portents of unusual events. But coincidences between events in the sky and those on Earth that were put forward as proofs of signs or even influence, were easy to dismiss because they lacked explanations. And in this case explanations that could be tested were painfully slow in forthcoming. In fact it took more than two hundred years from their first discovery before any regularities in the spots' appearances were detected. We shall see that, as luck would have it, soon afterwards a connection between sunspots and a measurable physical entity was discovered. As a result it came to be accepted that the Sun does exert influence on the Earth in ways our five senses are not able to register without the help of instruments. Since then sunspot studies and their influence on human life have again become respectable subjects of research.

Enthusiasm in the late seventeenth century could also have waned because Scheiner's *Rosa Ursina* was such a boring read and besides, it seemed to summarize everything about the spots that could ever have been known. To watch without any hope of a discovery was not very alluring.

Last but not least, there can not be any doubt now that in that same period sunspots were rather scarce. In fact none were seen for several years, so much so, that the appearance of a large spot in 1671 caused quite a sensation. Who would want to study sunspots if there are so few to be found?

I don't mean that no observations were made at all, but the momentum was noticeably weaker than immediately after the first discoveries were made and after the big fight for priority. It turned out that Robert Boyle followed the passage of a large spot in 1660 but he dug up his old notes about it only in 1671, after the Italian born astronomer Giovanni Domenico Cassini announced that he saw a large spot. Cassini, invited by Louis XIV to become director of the newly founded Paris Observatory, sparked a minor controversy by stating that previously no spots had been seen for twenty years. This was contradicted not only

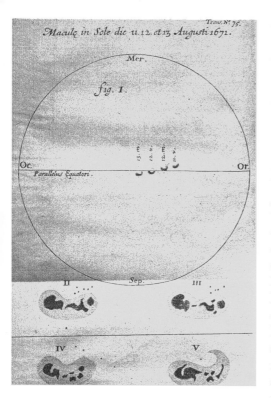

Maculæ in Sole die 11.12. et 13. Augusti 1671.

Trans. N° 75.

fig. 1.

Mer.

Oc.
Parallelus Æquatori.
Or.

Sep.

II   III

IV   V

Sunspot drawing from Philosophical Transactions, *1671.*

by Boyle but also by one of Cassini's colleagues, Jean Picard. Aboard ship, and because of unfavourable wind at anchor at the island of Vlieland off the Netherlands, Picard reported seeing the same spot as Cassini and mentioned another one he had seen ten years earlier.

John Flamsteed the first Astronomer Royal, appointed in 1675 with the brief to improve the star charts for the navigational tables, did have an interest in the Sun as well. He equipped a solar observatory at Greenwich and made calculations about the angle of the Sun's axis and about its speed of rotation. But Flamsteed's fame rests on the star catalogue he published and not on any solar observations. His story is worth telling. In spite of his frail health Flamsteed was a hard working perfectionist and as such reluctant to publish a less than perfect star catalogue. This did not please the members of the Royal Society, nor Newton, its president. After years of waiting Flamsteed was summoned. Newton was in a rage. Flamsteed recalled: he 'told me how much I had received from the Government in 36 years I had served. I asked what he had done for the £500 per annum that he had received ever since he had settled in London.' Newton 'called me all the ill names, puppy, etc, that he could think of.'[3] Newton's hostility may even have prevented sunspot observations being published in the *Philosophical Transactions,* the journal of the Royal Society, by correspondents of Flamsteed.

Flamsteed agreed to deposit a sealed copy of the incomplete data as his pledge to deliver more complete copy. He stipulated that nothing should be published until he was ready to supply further data. The Royal Society waited six years for the data to materialize. Finally they lost patience, broke the seal and in 1712

THE ENIGMA OF SUNSPOTS

published what they had on hand together with additional data by Halley. Flamsteed was angry beyond belief and burned all the copies of the publication he could lay his hands on, about 300 all told. Although this incident spurred him to work even harder than before, publication of his star catalogue, the best of the time, was only completed in 1725, six years after Flamsteed's death.

Flamsteed had a fleet of amateur correspondents who kept him informed about sunspots though he made his own observations as well. Many of his informers were clergymen, like William Derham, vicar at Upminster in Essex, afterwards chaplain to the Prince of Wales, later George II. Derham, we are told: 'lived quietly, cultivating his tastes for natural history and mechanics,'[4] taking notes of the weather, insects, migrating birds, and of course, the stars and the Sun. This did not prevent Derham

John Flamsteed.

from being an extremely successful lecturer and author, whose *Astrotheology* and *Physico-theology* was republished many times and translated into several languages. As a fellow of the Royal Society, Derham contributed to the *Philosophical Transactions*. He reckoned that sunspots were results of volcanic eruptions and fancied seeing smoke rising from them.

Derham and his fellow spot watchers were hindered by the scarcity of spots at the time. Later, when this state of affairs lapsed from consciousness, Derham was rather unfairly blamed for not observing regularly enough. But Flamsteed could see no spots between the years 1676 and 1683, or in the years between 1687 and 1694. He wrote to Derham about rumours circulating that spots were seen: 'As for spots in the Sun, there have been none since the year 1684' and 'All the stories you have heard of

*Aurora Borealis.
An engraving made
by the French Arctic
Expedition 1838–40.*

them are a silly romance spread [by] such as call themselves witty men to abuse the credulous.'[5] In 1711 Derham concluded that there were 'doubtless great intervals sometimes when the Sun is free ... Spots could hardly escape the sight of so many Observers of the Sun as were then perpetually peeping upon him with their Telescopes.'[6] At the time no great significance was attached to the dearth of spots. It could not have been, and was not, regarded as unusual because, as the astrophysicist Eugene Parker said: 'there had been 34 years with sunspots and 70 years without them. Who was to say what was normal?'[7] Hindsight tells us that during the time span of human history, seventy years without spots was definitely unusual.

After 1715 spots increased in number and size. Auroras (known as the Northern Lights or Aurora Borealis in the northern hemisphere, and Southern Lights or Aurora Australis in the southern hemisphere) also started to appear at low latitudes. In

THE ENIGMA OF SUNSPOTS

March 1716 spectacular Northern Lights were seen in London and all over Europe as far south as Italy. This magnificent display was greeted with 'mingled wonder, awe, and admiration.' The show started around 7 pm when Edmund Halley, who later gave a thorough description of it, was at a friend's house. It was a cold night and they must have sat with curtains or shutters drawn because news of the extraordinary happening only reached them two hours after it first started. Halley and his friend rushed to the window but this was in a southerly direction. Not wanting to miss anything of the spectacular they went outdoors and braved the cold. They watched until 11 pm when, frozen to the bone, Halley finally returned home and carried on watching from the comparative comfort of his house. The display did not cease before 3 am.

Impressed by what he saw, Halley tried to develop a magnetic theory of the auroras. He conceived of magnetism as:

> subtile Matter, [which] no otherways discovering
> itself but by its Effects on the Magnetic Needle, wholly
> imperceptible, and at other times invisible, may now
> and then by the Concourse of several causes very
> rarely coincident, and to us yet unknown, be capable
> of producing a small degree of Light[8]

either by becoming luminous itself or, carrying from the interior of the Earth something that could be responsible for the phenomenon. Halley was partially on the right track: magnetic activity is behind the appearance of the auroral display. But it is influenced primarily by the Sun's magnetic activity and not that of the Earth.

The Sun's magnetic activity governs the sunspots' appearances and that of the auroras as well. Actually that spots and auroras may somehow be connected has occurred to Dortois de Mairan in the 1750s. De Mairan compiled a historical list of auroras and found that neither spots nor auroras were numerous between the years 1620 and 1710. He speculated that auroras were caused by the solar atmosphere mingling with the terrestrial atmosphere. A good intuition, all it needs is the substitution

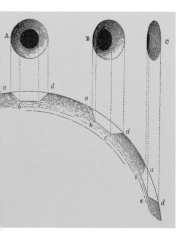

The change in appearance of a sunspot as it approaches the limb of the Sun. From Secchi, *Le Soleil, 1875*.

of the word 'magnetism' for 'atmosphere' and a more precise expression instead of 'mingling.'

For 150 years the nature of the spots was subject of pure speculation. The first breakthrough came in November 1769. During that month Alexander Wilson observed a huge spot moving across the face of the Sun and noticed how it changed when approaching the limb. His interpretation of the changes this spot underwent is still valid today.

Born in 1714 in St Andrews in Scotland, Wilson sampled several careers as a surgeon and apothecary and as a type founder before turning to astronomy. But even at an early age he spent much of his free time experimenting with the scientific instruments he constructed for himself. 'Possessing naturally much activity of mind, and employing most of his leisure in some ingenious attempt or other' — as his son put it — Wilson conducted atmospheric temperature measurements with a kite.[9] Luckily for him and also for astronomy, there were no thunderstorms at the time, otherwise he might have been killed by lightning. Wilson gained so much expertise in scientific matters that in 1760 he was appointed to the newly established chair of astronomy in Glasgow.

One of Wilson's friends wrote to him from London about the appearance of an unusually large spot, but being cloudy for several days he was unable to see it. When the Sun finally appeared on November 22, 1769, the spot was already near its edge. Wilson was lucky in that the weather was calm and he could see clearly without any atmospheric disturbance. Using a telescope with a 'glass properly smoked,' not a camera obscura, the next day he found that the penumbra (the not-so-dark region surrounding the nucleus, or umbra) contracted on the side that was further away from the edge. This was exactly the opposite of what he expected to see. Next day the penumbra on that side disappeared completely and even the nucleus seemed slimmer.

Wilson was getting more and more dumbfounded. Trying to explain what happened he considered two possibilities. One was

THE ENIGMA OF SUNSPOTS

that the spot had changed. This was not unusual; spots often change their shape. The second possibility was that the appearance had something to do with the Sun's rotation. Wilson hit on the idea that the observed changes could be explained by envisaging the spots as indentations (he called them 'excavations') on the surface of the Sun. To make sure that he got the perspective correctly, Wilson made a white globe and drilled small holes into it. He painted the bottoms of the holes black and inspected what happened when he looked at them from afar while the globe slowly turned on a metal axis. He was satisfied that the model replicated the spot's behaviour.

The model was correct but was he right about the sunspots? If he was, then the part of the penumbra that disappeared from sight did not vanish in reality and was hidden from view only because it was sloping away. Wilson was hoping for the same spot to come back again as it rotated with the Sun. It did return on December 11. If the spot itself had changed, then this time the penumbra closer to the edge would have been thinner. But Wilson found that when the spot first appeared at the edge of the Sun, it did not have a penumbra at all on the side closer to the centre. The next day, as it gradually moved towards the middle of the disc the penumbra became visible but it was still thinner than the penumbra on the other side. We can imagine Wilson's frustration when for the next five days it was cloudy again and he had to abandon his observation. Luck returned on December 17 when the weather cleared. By then the spot was closer to the middle of the Sun and he could see it again completely surrounded by the penumbra.

One single observation of course, would not be enough to formulate an acceptable theory. Wilson carried on watching other spots and their pattern of behaviour turned out to be similar. As we shall see, our present understanding still is that spots are indentations but not in the sense that Wilson imagined.

About previous speculations Wilson said: 'Conjectures concerning the nature of the sun, were early indulged in,' but 'every theory, how ingenious soever, which is founded upon a misapprehension of things is apt to be pressed with many difficulties.'[10]

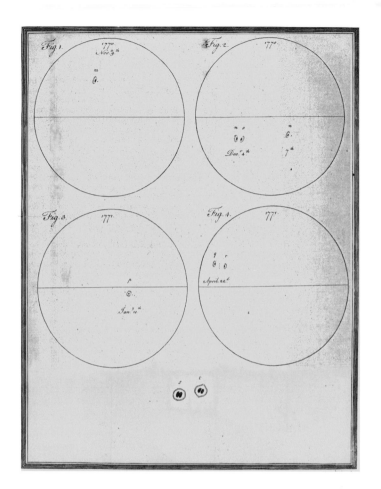

Drawings of the Wilson effect 1770. From Philosophical Transactions *, Vol. XIV.*

A polite but true remark. It did not however, prevent Wilson from adding his own to the mountain of already existing speculations. That he phrased his own conjectures in a series of queries instead of statements, a method frequently used by scientists, does not change the fact that they are but conjectures and as such 'apt to be pressed with many difficulties.'

Wilson conjectured that the Sun was a dark globe with an irregular surface surrounded by a luminous envelope. According to him sunspots are produced in the following way: a gas rises

96   *THE ENIGMA OF SUNSPOTS*

from the interior, it expands and tears the outside layer apart, uncovering the dark body of the Sun. These holes in the outside layer are the spots. The outside luminous layer can not be a liquid or a gas because that would immediately fill the holes. It must have the consistency of a thick fog that slowly and gradually fills up the holes and makes the spots disappear.

Wilson's theory on the nature of the Sun and its spots did not go uncontested. Samuel Dunn, a teacher of mathematics and a prolific author on navigation, who claimed that he discovered the strange behaviour of the penumbra several years before Wilson, proposed a neat idea that contradicted the indentations. He thought that the penumbra might be a ring around the spots, much like the ring around the planet Saturn. The spots, hovering inside this ring occasionally cover part of it. Did this explain the phenomena? It did, superficially. But why would the spot obscure part of the ring every time it approaches the edge? Dunn's explanation could not answer this simple question while Wilson's could. As a result Dunn and his theory have been conveniently forgotten.

It was Jerome Lalande who criticized Wilson's ideas especially vigorously. A master of self-promotion and an excellent popularizer of astronomy, Lalande was director of the Paris Observatory. He was notorious for causing panic in Paris in 1773 when he discussed a collision between the Earth and a comet but did not disclose how unlikely such an event might be.

Lalande was no stranger to controversy either, and seemed to revel in scientific argumentation. Many of the spots he observed did not behave the way those seen by Wilson did. Therefore sunspots were not holes in the covering layer of the Sun through which we can look into an abyss, but on the contrary, high mountain peaks occasionally uncovered by the ebbing of a luminous ocean. Lalande objected to Wilson's theory based on what he termed 'equivocal' observations. In his own defence Wilson — who was obviously unhappy about Lalande's attitude — pointed out that some spots might just be too shallow for the effect to be seen unequivocally. He drew attention to the immense scale difference between the Sun's radius and the depth of the indentation. He added rather unkindly that Lalande's 'clear and comprehensive ideas of everything ... have doubtless led him to think that any particular attention to exactness was unnecessary.'[11]

Actually, the main argument between Lalande and Wilson centred on a philosophical question: namely when is a theory based on and deduced from observation or experiment, and when is it pure conjecture? Wilson maintained that he effectively separated the two and his assertion that spots were indentations was based on valid observations. Lalande attacked the rest of the conjectures as if Wilson pretended to have deduced those as well.

A 'fatal objection' — in the words of a nineteenth-century historian — to both Wilson's and Lalande's theories is, that they do not explain the sharp dividing line between the penumbra and surrounding luminous region. To be honest, neither theory was brand new. These notions had been floating around for at least a hundred years already. The old theories, now resurrected to explain new observations were not entirely successful and induced the German astronomer J.E. Bode to make some improvements. He added an extra layer surrounding the Sun. One of the layers would consist of an opaque, vapour-like substance and the other one would be a luminous outer layer. Where the luminous layer is thinner we see the penumbra, where it has a proper hole we see the dark, lower atmosphere. At places where the luminous layer heaves up, we can see the faculae.

The existence of sunspots certainly gave people the opportunity to carry on making various conjectures about the Sun. Human inventiveness and imagination can be practically limitless if information is missing or is misinterpreted. The case of Charles Palmer (an accountant by profession) shows how easy it is to arrive at ridiculous theories based on real experience and fairly rational thought. He seriously suggested that the Sun is a convex lens made of ice that gathers the light of the sky and reflects the heat of the Earth onto itself. Palmer was not completely out of his mind. He reasoned that light could not possibly originate from the Sun but must come from the heavens, because according to the Bible God created light on the first day and the Sun only on the fourth day. Furthermore, if the Sun was the source of heat then the higher up you went, the closer you were

THE ENIGMA OF SUNSPOTS

to it and the warmer it should get. But experience shows that on top of high mountains it is much colder than lower down. From this it follows that the Sun cannot be hot. Palmer would have been absolutely right if the heat of the Sun reached us either by conduction or by convection. But the heat of the Sun arrives here by radiation, a method of heat transmission he did not know about because Sir William Herschel discovered it only after Palmer published his theory in 1798.

Why did Palmer think the Sun was made of ice? Convex lenses collect light and they are round, they are also transparent. As a material Palmer had to discount glass because it is man made and he lived much too early to have heard about plastics. He could have chosen rock crystal but a lens made of ice was much more likely. When we follow Palmer's line of thinking it does not sound all that crazy.

Palmer thought the Sun was a block of ice. The Reverend Tobias Swinden knew better: he knew that it was hot, very hot. Swinden was modest enough to declare that the 'particular Description of the Sun's Body so as to set down the Modus of its Existence or Operation'[12] was a task too difficult for him to undertake. But in 1714 he managed to prove beyond a doubt that the Sun was nothing else but hell itself where fallen angels and damned souls congregate. He dismissed ideas that hell might be in the centre of the Earth because no fire could last long enough there without air and anyway the Earth would be far too small to accommodate all the damned, past, present and future. Swinden decided that sunspots were the 'Seats of the Blackness of Darkness' where the wicked are perpetually deprived of light.

In 1755 the philosopher Immanuel Kant described the Sun as a 'flaming globe,' a 'mass of molten glowing matter,' with a dark core. On the outer layer are:

> wide seas of fire ... raging storms ... which they swell over their shores, now cover the elevated regions of this celestial body, now sink back into their confines; burnt out rocks which stick out their fearful peaks from the flaming gorges and whose inundation or unveiling by the waving fire

elements causes the alternating appearance and disappearance of sunspots.[13]

The translator, S. Jaki remarks with a slight understatement: 'The ease with which Kant discourses about deep gorges inside the Sun is worth noting.'[14]

In 1775 J.E.B. Wiedeburg, professor of theology and mathematics at Jena, combined the two popular opinions on comets and sunspots. Setting himself the task of reconciling science with the words of the Bible, he had to account for the creation of the Earth. (Wiedeburg was not alone grappling with this problem.) This is what he came up with. The Sun, in common with the other heavenly bodies, constantly ejects tiny, very nearly weightless particles. When these clump together we see them as spots which therefore are not on the Sun itself but extremely close to it. Heavier bits sink into the centre of this conglomeration, condense and form the nucleus of a sunspot. Some spots fall back into the Sun but some are repelled by the Sun's electrical force and become comets and planets. That is how the Earth was made. In fact the Earth is nothing else but an old sunspot. Wiedeburg conceded that his theory had its problems. How come we do not notice the Sun shrinking if it constantly loses material? By analogy he reminds us that perspiration is imperceptible and we do not notice losing any material by it, neither does the sea shrink although it constantly evaporates. Wiedeburg did not seem to notice that we drink in order to replenish the lost fluid and that the sea is also replenished all the time. Where are the new planets and comets that are formed? The answer to this was that there might be undiscovered new planets out there.

Herschel's Strange Ideas

The famous astronomer, discoverer of the planet Uranus, Sir William Herschel also joined the band of speculators. Herschel started out as a musician in Hanover, like his father before him. He came to England in 1757 as a refugee (more like a deserter) from the Seven Year War and settled in Bath as an organist and music teacher. Soon he brought his sister Caroline over from Germany to help him. Caroline Herschel kept house for her brother and helped him not only with his music but also with his astronomy. In fact she became an active and successful astronomer in her own right, discoverer of several comets and nebulae. The Royal Astronomical Society recognized Caroline Herschel's work, awarded her a gold medal and made her an honorary member.

At first, astronomy was just one of Herschel's hobbies and he started polishing his own lenses and casting and polishing his mirrors. He used mostly reflecting telescopes and had to make his own because he could not afford to buy a really good one. In the beginning it took two hundred attempts before he was satisfied with the quality of a mirror. Meanwhile the whole house was turned into a workshop with a huge lathe occupying the middle of the bedroom and a cabinetmaker working in the drawing room.

When George III appointed Herschel his private astronomer and Caroline the official assistant, brother and sister moved to Slough, close to the Royal residence in Windsor. Now the Herschels could stop giving music lessons, but they still had to supplement their income with making and selling telescopes. Soon however, their financial problems came to an end when Herschel married a rich widow.

Sir William Herschel.

At first Herschel used coloured glasses when watching the Sun. Later he tried using coloured liquids, but the heat of the rays caused turbulence and this impaired the vision. Not wanting to stop down the aperture of his lenses Herschel then used the method Hooke proposed a hundred years earlier and split the rays by repeated reflection.

Like Wilson and Bode, Herschel also held that the interior of the Sun was cool and dark. Unlike Wilson and Bode, Herschel held that the Sun was inhabited. He also believed that sunspots influence the weather here on Earth. A recent biographer was charitable: Herschel 'was not without an occasional peculiar idea.'[1] Less charitable were the contemporaries and some of the earlier critics. One reviewer could not envisage any causal connection and declared Herschel's a 'hasty and erroneous theory concerning the influence of solar spots on the price of grain. Since the publication of Gulliver's voyages to Laputa, nothing so ridiculous has been offered to the world.'[2] 'Posterity will rank this opinion on the Sun among the aberrations of the human mind' was another comment.[3]

Although the idea that the planets were inhabited was fairly widespread, belief in an inhabited Sun was not. Huygens, who speculated extensively about the inhabitants of the various planets and about their physical and mental characteristics, stopped short of ascribing any to the Sun. In fact, belief in an inhabited Sun was reason enough to be considered a lunatic as the case of a certain Dr Elliot shows. In 1787 this gentleman fired pistol shots at a couple taking a peaceful walk. Elliot was acquitted of attempted murder because no bullets were found at the scene and he was charged with the lesser crime of assault. The case of the defence was insanity on the grounds that Elliot held the crazy belief that the light of the Sun:

THE ENIGMA OF SUNSPOTS

Proceeds from a dense and universal aurora, which may afford ample light to the inhabitants of the surface (of the Sun) beneath ... there may be water and dry land, hills and dales, rain and fair weather; and as the light, so the season must be eternal; consequently it may easily be conceived to be far the most blissful habitation of the whole system.[4]

We shall never know what the jury and the judge made of the case because Dr Elliot starved himself to death in Newgate Prison while waiting for the trial to end.

Unlike Dr Elliot, Herschel was never accused of insanity for his beliefs. It has to be admitted that they were only beliefs, based on philosophical and not on scientific considerations. Herschel could not accept that the Sun's only function was to provide us with heat and light, he thought it would be a 'wasted' globe if it was not inhabited. Building on Wilson's finding that the spots are indentations, Herschel's vision floated two regions of clouds in the atmosphere above the inhabited dark sphere of the Sun. This scenario was much like Bode's. The lower layer is opaque, it shields the dark interior from the hot and luminous upper layer and at the same time it reflects its light. When both cloud layers have an equal hole, then the dark Sun is exposed and we see the nucleus of a sunspot. When the opening in the upper clouds is larger than the opening in the lower ones, then we see a spot surrounded by a penumbra. If only the lower clouds are parted, then we see a penumbra without a nucleus. Although this was a better explanation than Wilson's, it also required a gas to escape from the interior and punch holes in the cloud cover.

For this there was no experimental evidence and a critique given to Wilson's theory applies equally to Herschel's: 'Speculations ... as they involve a gaseous vapour, of whose existence we have no proof ... are inadmissible into the rank of physical truths deduced by legitimate reasoning from established facts.'[5]

Herschel tried to introduce a new terminology, in fact he was accused of 'idle fondness for inventing names without any manner of occasion.' He called the nuclei of the spots 'openings,' the *faculae* 'nodules' and the various other patterns on the face of the

Sun he named 'shallows, ridges, corrugations, indentations and pores.' Herschel was the first to describe filaments seen in the penumbra: he called them 'tufted shallows.' This new nomenclature never caught on, partly because it reflected the structure of the Sun as Herschel imagined it and partly because adequate names for the various features already existed.

Of more lasting influence were Herschel's investigations about the terrestrial influence of solar activity. We have already seen that the Greeks and the Romans coupled sunspots with wind and rain. In 1651 Riccioli, the astronomer who named many of the lunar craters, said that an abundance of spots is associated with cold and rainy weather. So did Kircher in 1671 and Wiedeburg in 1729. But Herschel was the first to present statistical, not just anecdotal, evidence. Comparing the price of wheat in Windsor between the years 1650 and 1713 with the number of sunspots, he found that on the whole when sunspots were rare wheat was more expensive than in the years when more spots were seen. Dear wheat prices meant poor harvest of course, in all probability the result of a cloudy and cool summer. The years between 1795 and 1800 also produced poor harvests and few spots.

Before discussing any further the possible wheat-price/sunspot connection, we have to distinguish carefully between Herschel's notion of influence and the kind of influence some of those who believe in astrology are thinking of. This conceptual difference regularly surfaces, unfortunately without any understanding being reached between the two different mindsets. There can be no doubt that Herschel assumed the existence of a physical mechanism, which caused warmer summers when spots were plentiful. He did not know what this mechanism might be, but trusted that sooner or later it could be found. Whether his conjectures were correct or not could also be checked; he himself tried to do so. On the other hand there were, and still are, those who think that a physical mechanism cannot and at any rate need not be found, because a sympathetic (or mystical?) connection exists between the macrocosm and the microcosm, between the universe at large, and human affairs.

How did Herschel explain his findings about wheat prices? Spots block the surface of the Sun, so intuitively the weather should be colder when there are many of them. But if for some reason the spots were warmer than the rest of the surface, then that would account for cooler summers when there are fewer spots and warmer summers when there are many. Herschel himself was not very happy with this argument, partly because it starkly contradicted his own weather observations, and partly because it required the dark — according to him inhabited — centre of the Sun to radiate more heat than the surface. Surely, such a body cannot be solid and certainly it could not sustain life as we know it.

Then in 1845 Joseph Henry, an American professor, came to show that sunspots radiate less heat than their surrounding area. His method made use of a recently invented measuring instrument, the 'thermocouple.' If we make a circuit using wires of two different metals, such as for instance copper and iron, and heat one of the junctions where the two wires meet, an electric current will flow. The warmer the junction of the two metals is, the more current will flow. Actually, the current produced like that is rather small, so small that it is hard to measure. A thermopile, which is nothing else but several thermocouples connected together, gives currents that are easier to detect. This is what Henry used.

Henry stuck a telescope through a window in the hallway of his house and projected the picture of the Sun onto a screen just like other sunspot watchers before him had done. He made a hole in the screen for the light (and radiant heat) of an exceptionally large sunspot (it was more than 10,000 miles in diameter) to pass through. He put a thermopile behind the screen and let only the rays from the spot fall on it. He measured the strength of the current produced. Then he moved the screen slightly, to catch a bright part of the Sun through the hole and again measured the current. The heat from the spot area produced less current than the brighter area, so it was cooler. Note that he measured the relative temperature between the two areas and not the actual temperature of the spot. We cannot quibble with Henry's measurements: sunspots are cooler, not warmer, than the brighter areas.* One flaw in Herschel's argument is that the price of wheat in Windsor does not depend exclusively on the summer temperature. It depends on the rainfall and it also depends on the political situation, to name but two additional factors. Correlation of such short duration could also be due to chance coincidence.

That Herschel's conclusion was hasty and not well founded seemed all the more likely as several other observers, trying to verify his hypothesis, could not find any connection between the number of sunspots and the temperature. Indeed, Alfred Gautier of the observatory in Geneva came up with a diametrically opposite correlation. Gautier studied sunspot records from the first half of the nineteenth century and when he compared them with ten year averages of temperature records from Paris, Geneva and the Great St Bernard Pass, he found in complete contrast to Herschel that years rich in spots were in effect cooler.

Was Herschel entirely wrong? Was Gautier wrong? As I am writing this, the question has still not been finally resolved. Historical reconstruction shows that at certain periods the correlation between temperature and the Sun's activity is probably positive, while at certain other periods it might be negative. This leads us to the conclusion that both Herschel and Gautier might have been correct. Herschel published his findings in 1801, relying on seventeenth- and eighteenth-century data when lots of sunspots were accompanied by warm weather, while in the nineteenth century which supplied Gautier's figures, the situation was exactly the opposite. Solar physicists believe that right now the hotter areas, such as the faculae, more than compensate for the heat lost by the cool spots. Finally, if we look carefully, we can see that most of Herschel's data came from an anomalous period in the history of the Sun. According to the American astrophysicist H.T. Stetson, it is possible that under a certain critical value of the number of spots the temperature rises with the increase of spots and over that critical value it falls when the spots increase. Since in the period Herschel examined the number of spots were at a historical low, they must have been under this hypothetical

* The temperature of a sunspot nucleus is about 4240°K while the mean surface temperature of the Sun is 5780°K.

THE ENIGMA OF SUNSPOTS

Dominique François Jean Arago.

critical value. Therefore it is feasible that the temperature was higher when spots were more numerous.

Not everyone rejected Herschel's ideas as vigorously as the quotations above might indicate. Arago, the successor of Lalande as director of the Paris Observatory, was enthusiastic. (But then Arago was enthusiastic about a great many things.) His interest, awakened by Herschel, breathed new life into the ancient notion that the weather and the sunspots are somehow connected. Arago kept regular sunspot and weather observations, published them in his journal, and urged others all over the world to do the same. Globalization began in earnest. And when the Baron Alexander Humboldt called for a worldwide survey of terrestrial magnetism in the 1830s, then it really took off.

Arago and his good friend Humboldt made many important and lasting contributions to science and were very influential in their time. Indeed, it was held that: 'with the exception of Napoleon Bonaparte, [Humboldt] was the most famous man in Europe.' Yet nowadays neither Arago nor Humboldt are names that come easily to mind when we think of famous scientists. The reason for this could be found in their spreading their talents far and wide, probably too far and too wide, and as a consequence no laws or principles bearing their names are taught in school science classes. They were both adventurers, but while Humboldt looked for it, adventure came to Arago uninvited.

In 1806 the French Bureau of Longitude sent Arago to Spain to finish surveying the line of longitude that runs through Paris. This was the time of the Napoleonic wars when French troops occupied part of Spain and engendered the hostility of the population against the French. As Isaac Asimov says: Arago's 'surveys involved hairbreadth escapes that would have read well in a thriller.' On occasions Arago and his colleagues had the protection of a bandit chief to whom they had once given shelter under the illusion that he was a customs officer. Sometimes Arago had to run for his life and once at night, when he was pursued by an angry crowd, in desperation he knocked at a peasant's house who luckily was willing to give him refuge. For a while on the island of Majorca he had to hide on a small boat in disguise. Then he travelled to Algiers on a fishing vessel, hoping to make his way back to France from there. But his false passport as an itinerant Hungarian merchant was not much use when the boat was captured by Spanish privateers. The passengers, together with the monkeys and the two lions the boat was also carrying as presents from the governor of Algiers to Napoleon, were taken back to Spain. Arago landed in prison. His jailers looked at the surveying notes he had in his pockets and took them for secret sign writing. His fluent Spanish was not in his favour either, so in spite of

THE ENIGMA OF SUNSPOTS

all his protestations they suspected him of being a spy. Arago spent the next three months in prison and even had to sell his watch to buy food. By a strange coincidence the watch turned up in France, was recognized as Arago's and he was believed to be dead. One of the lions really died in captivity and it took the threat of war by Algiers — because of the surviving lion, not because of Arago — to set the prisoners free. It took Arago a whole year to get back to Paris, alive, to the pleasant surprise of the folks at home.

By contrast, Humboldt *did* go looking for adventure and in 1799 he went off to explore the Americas. During the five years he spent travelling in South and Central America, he made a holistic study of the area. He studied the plants and the animals, the geology, the rivers, the oceans, the volcanoes, the stars of the southern hemisphere, and the cultural life of the inhabitants. Among other daring feats, he explored the Orinoco and Amazon rivers and climbed Mount Chimborazo. He also took magnetic measurements wherever he went and found that the field strength varied from place to place.

Ever since the invention of the compass and its use for navigational purposes it had always been important to know how certain magnetic elements varied in time and by geographical location. Already Kircher had asked his friends all over Europe to send him information about the movement of the magnetic needle. When the voyages of discovery got under way, measurements were taken wherever the adventurers went. But now Humboldt became convinced that the mapping of the Earth's magnetic field was not only of practical, but also of great scientific importance. On his return from America he initiated international cooperation for the study of terrestrial magnetism. During a trip to Siberia he successfully suggested to the Russian government to set up a line of magnetic and meteorological stations and in 1836 he managed to persuade the British Government to install magnetic observatories in the colonies. At Humboldt's recommendation the East India Company embarked on the geomagnetic survey of India in the 1850s. Humboldt's influence led Karl Friedrich Gauss, the 'prince of mathematics,' to organize the German Magnetic Union whereby observations were taken at a fixed time at several different places in Europe and collected together. (Humboldt boasted that he had awakened Gauss' interest in magnetism, but was less than politely enlightened when Gauss declared he had already been interested in the subject for the last thirty years.) Thanks to Humboldt the study of terrestrial magnetism became fashionable and this in turn had a revolutionary influence on the study of sunspots.

Monitoring geomagnetism is more important now than ever before. Currently all over the world, though mostly in the northern hemisphere, over 200 terrestrial magnetic observatories take constant measurements and several spacecraft are doing the same. But what has magnetism to do with sunspots? We shall soon find out.

∿ 12 ↩

Matching Cycles

Only a genius in the class of Galileo or Newton can discover anything important in science. True? Not so! In the diligent German amateur Heinrich Schwabe we meet a counter example to this mistaken opinion. Simple perseverance and attention to detail led him to what has become one of the cornerstones of solar and solar-terrestrial research. Schwabe was a pharmacist by trade but as soon as he prudently could, he sold his shop and having sufficient funds, devoted his time to astronomy.

When Schwabe started his observations it was in the hope that he would be the lucky discoverer of a new planet, but soon he became completely besotted with sunspots instead. For about forty years Schwabe watched the Sun daily whenever possible and averaged an annual 300 days of observations. Sunspots were then regarded as mere curiosities without displaying any of the regularities so important for science. But Schwabe, whose vigil started in 1826, could by 1844 discern a certain periodicity in the number of spots appearing. For about five years the number of spots declined to reach a definite minimum, but the next five years or so, their number increased and reached a definite maximum. Then the whole process started all over again.

Although several astronomers made use of Schwabe's data, his discovery of the cyclical nature of spots was not greeted with great hullabaloo. By the middle of the nineteenth century interest in sunspots was again at a low ebb after the enthusiasm engendered by Herschel subsided. Rudolf Wolf, director of the observatory in Berne, and afterwards in Zürich in Switzerland, attributed the lack of interest in Schwabe's announcement to plain prejudice, in other words, to the usual difficulty of having new ideas accepted. There may be some truth in this, after all, for two hundred years sunspots were thought to be entirely haphazard features and it was difficult to suddenly digest that they displayed any regularity. But fixed attitudes were only part of the problem; the data presented its own problems as well, namely the difficulty of quantifying sunspots. To start with, however regularly Schwabe may have looked, he certainly could not make any observations on days when the Sun was obscured by clouds and as a result there were gaps in his data. Then, he took into account only spot groups and ignored individual spots however large they may have been. He counted as a group two or more spots if there was no space between their penumbra. But this depends very much on the instruments used. A telescope with greater power might show gaps between penumbras where one with less resolution might not. Last but not least, can one honestly accept a 10-year-cycle when data of only a couple of presumed cycles is available?

Had it not been for Wolf's extraordinary effort and an additional, coincidental discovery, it is quite possible that the periodicity, and especially the length of the supposed period of the sunspots would not have been generally accepted for a long time. Wolf was another one of those who, once seeing a large group became instantly obsessed with sunspots. From 1848 onwards he could not let a day go by without at least trying to watch the Sun. At first he could not make much with Schwabe's announcement either, but then suddenly a connection popped up where none existed before. This was magnetism.

Remember Humboldt and his drive to take magnetic readings all over the globe? Humboldt was one of the last men who managed to know nearly everything about nearly everything. Between 1845 and 1858 he published several volumes of a comprehensive work, *The Cosmos*. As its title implies it was wide-ranging, to say the least. Among the astronomical, geophysical and geographical subjects he discussed both terrestrial magnetism and recent progress in astronomy, and drew public attention to Schwabe's work. Were minds triggered by Humboldt's disclosure, or was it a coincidence that within a short time three independent announcements linked changes in the

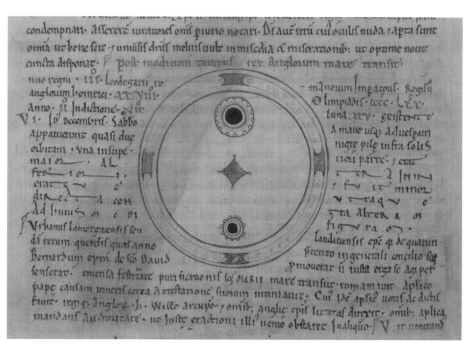

Above: MS CCC 157 p.380 Sunspots seen in the reign of Henry I, inserted c.1140 by John of Worcester into the chronicle begun by Florence of Worcester (d.1118) c.1130-40 (parchment) by English School (12th century), Worcester Chronicle, (12th century), Corpus Christi College, Oxford, UK / Bridgeman Art Library.

Below: Photograph of the Sun taken in 2001, showing the scale of the Earth. (Nasa)

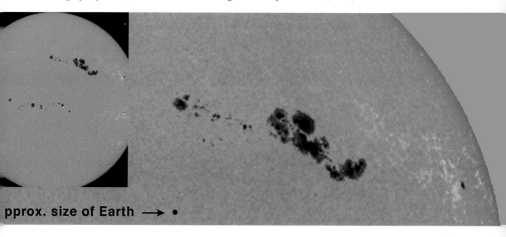

pprox. size of Earth ⟶ •

Earth's magnetic field to Schwabe's sunspot cycle? *The Cosmos* covered so many diverse topics that it could have helped to uncover hidden connections. After all, this is what creative thinking mostly consists of: making linkages where none existed before. The president of the Royal Astronomical Society when honouring Schwabe with the Society's gold medal expressed the opinion that the publication of *The Cosmos* brought attention to the cycles. This has been repeated many times in the literature but I am not so sure that it is true. Several astronomers, among them Gautier for instance, confessed that hearing of Schwabe's discovery prompted them to take the study of sunspots seriously. It just took some time to draw the threads together. Let's see what these threads were.

The compass needle points to the Earth's magnetic north and south poles. The magnetic poles do not lie exactly on the geographical poles, so the needle does not line up with the meridian (the line of longitude running from geographic pole to pole). The difference in direction is called the 'declination.' To complicate matters even further, the compass needle does not stay still either, but keeps changing its position. Sometimes it points a little bit to the east, then again a little bit to the west. Some of these changes are slow and happen over hundreds of years. Some changes are violently fast, then we can say that we have a magnetic storm.

However, some changes occur every day. In the northern hemisphere the compass needle points a little to the northeast in the morning and a little to the northwest in the afternoon. The biggest change the needle makes from the declination is called the 'variation' for that day. In 1852 Johann Lamont, a Scottish astronomer working at the observatory near Munich reviewed magnetic observations from 1835 to 1850 and found that the range of daily variation had a cycle of just over ten years. For about five years or so, it decreased and reached a minimum, for the next five years or so it increased until it reached a maximum. Then the process started all over again. Familiar story? Not for Lamont, who was not much interested in the Sun and did not notice the strange coincidence between the periodic change in sunspot numbers and the cycles in the changes of the magnetic field.

What Lamont did not notice, others did. For the second time in our story we meet with three men claiming priority for a discovery. It was not long before Edward Sabine, Alfred Gautier and Rudolf Wolf, independently from each other, declared a connection between sunspot cycles and magnetic disturbances. They noticed not only that the two cycles were of the same duration, but that the maxima and minima fell together as well. This was truly too strange to be pure coincidence. Or was it? In Wolf's words: 'This correspondence between two phenomena, one of which seemed until now to belong exclusively to the Sun and the other exclusively to the Earth, was extraordinarily remarkable'[1] Wolf found this correspondence so remarkable that he immediately informed Humboldt, Faraday and the French Academy of it.

About the three men who discovered the same thing at the same time the good-natured Wolf said: 'The new Fabricius was not ignored by the new Galileo and Scheiner ... all three were pleased by the coincidence and by the result thus gained by science.'[2] Wolf was a little too magnanimous; the truth is that Sabine did quarrel both with him and with Lamont about the priority of the discovery. Sabine's legitimate point was that he noticed cycles in magnetic disturbances (storms) — in Toronto and in Hobart, Tasmania — and not in the cycles of variation, in other words that he did not base his discovery on Lamont's data as Wolf and Gautier did. For Sabine the magnetic disturbances were:

apparently indicating the existence of a periodical variation, which, either from a real or causal connection, or by a singular coincidence, corresponds precisely, both in period and epoch, with the variation in the frequency and magnitude of the solar spots, recently announced by M. Schwabe.[3]

Note how carefully Sabine chose his words: the correspondence could be a coincidence, but it could also be 'a real or causal'

Opposite:
*Visible light image
of Sun showing
sunspots.
(Science Photo
Library)*

This page:
*Solar flares
erupting from
the Sun's surface.
(U.S. National Solar
Observatory,
Sacramento Peak)*

connection (did he mean: either real, or causal, or did he mean: if it is real then it has to be causal?) he was unwilling to speculate about. He did however, make a tentative suggestion:

> In our present ignorance of the physical agency by which the periodical magnetic variations are produced, the possibility of the discovery of some cosmical connection ... should not be altogether overlooked.[4]

A colourful and varied career saw Sabine as a general in the Napoleonic Wars, and later an Arctic explorer with the Parry expedition. He worked for nearly 30 years at the Kew Observatory during which time he travelled all over the globe in his attempt to determine the shape of the Earth and its magnetic properties. Sabine was the mover and shaker behind the 'magnetic crusade': the mapping of the Earth's magnetic field. By contrast, Gautier and Wolf were Swiss astronomers, city dwellers whose interest was foremost in things astronomical, more in the Sun than in terrestrial magnetism.

Of the three original discoverers of the connection between sunspot cycles and cycles of magnetic variation, it fell to Rudolf Wolf to work out the details. Wolf knew what his strength was; it was mathematics. He also knew that he was no genius and set out to prove that with application even those with limited talents can achieve lasting results. He definitely succeeded. The possible connection between geomagnetism and sunspots intrigued him and gave him an added incentive to show that the cycles really existed.

Wolf concentrated on proving the existence of the sunspot cycle because data for terrestrial magnetism was still scarce and more unreliable than that for sunspots. In 1848 he had already started organizing the European observatories to record sunspots regularly and to do so in a standard way. By 1852 he had a four year supply of reliable data. To find out about spots in earlier times Wolf used the records of dozens of spot watchers, both those who were famous and those who were completely unknown within the scientific community. He went to libraries and archives and scoured literally hundreds of volumes of published and unpub-

THE ENIGMA OF SUNSPOTS

lished data on the spots. He wrote to astronomers asking them to do the same in places he could not visit personally. He even asked Richard Carrington, an English amateur astronomer to travel down to Petworth and examine Harriot's papers. Carrington faithfully copied Harriot's drawings and notes, and sent them to Wolf. The Petworth papers were not seriously studied again for another hundred years.

All in all, Wolf managed to gather information about sunspots for about 22,500 days, quite an achievement. Then he established a kind of an average for the spot numbers. He counted how many groups could be seen in a day, multiplied this by ten and added to it the number of individual spots. He averaged this out for months and for years and called it the 'relative sunspot number.' As time went by, he changed the formula slightly to account for the differences between individual observers.[5] The daily number can vary from zero to over two hundred. The current daily number together with all the daily numbers from 1818 to the present can be found on websites.[6]

By 1852 Wolf could present the average annual values of spots observed, going right back to the seventeenth century. Of course the older values were much less reliable than the more recent ones. Wolf then used a statistical method to establish the length of the sunspot cycle, which he calculated as just over eleven years. (This method is now a textbook example.) Wolf was aware that some cycles are shorter, some are longer, in fact they vary from 8 years to 17 years. The cycles are not symmetrical either, it takes longer to get from a maximum to a minimum than from a minimum to a maximum, but Wolf's result is fairly close to the value now accepted as the average length of the cycles.

Lately a couple of solar scientists, D.V. Hoyt and K.H. Schatten have done even better than Wolf in finding and evaluating old sunspot observations. True, they used modern computing methods but with application comparable to his, they rooted out journals and books Wolf missed and manuscripts he had not seen, doubling his already considerable number of observations. They reintroduced Schwabe's original method of

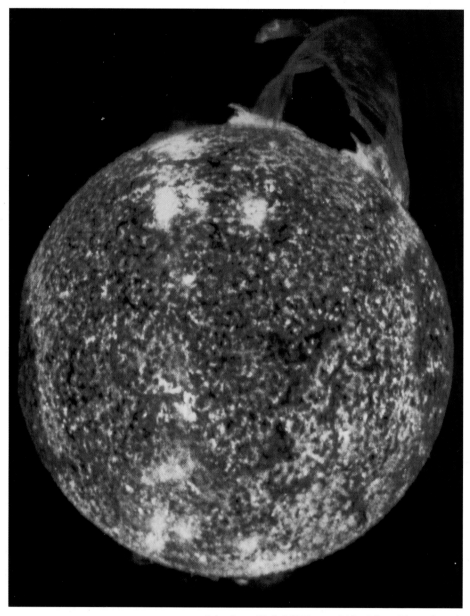

Solar flares. (Left: Still Pictures. Right: Science Photo Library).

THE ENIGMA OF SUNSPOTS

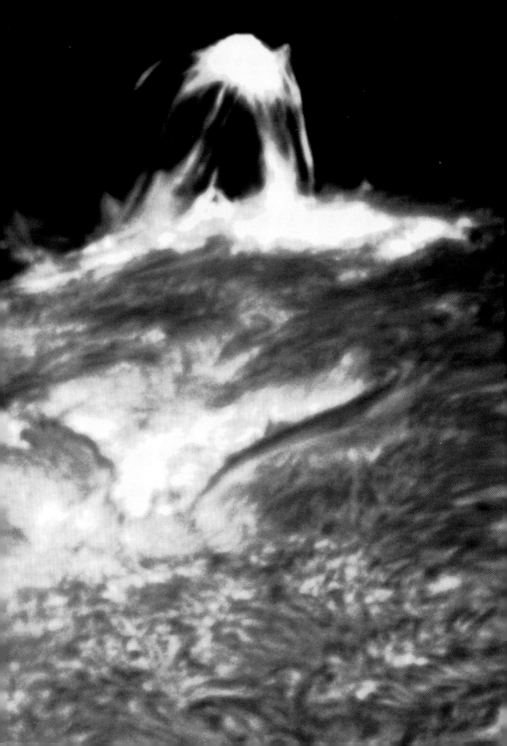

counting spot groups only. Also, they found some mistakes in Wolf's work, for instance one of Wolf's correspondents mixed up the number of groups with the number of individual spots. As a result, Hoyt and Schatten made adjustments to Wolf's calculations.

Sunspot cycles are accepted facts now, but not everybody believed Wolf when he announced his cycles. Although the Astronomical Society honoured Schwabe with its gold medal in 1857, some still doubted the existence of any periodicity while others were unhappy with the accuracy of the length of the cycle Wolf claimed to have established. Statistics were still in their infancy but people may have already been thinking along the lines of 'lies, damn lies and statistics.' Even Lamont remained unconvinced. Wolf had to search for more and more old observations trying to prove that he was right. Luckily he lived long enough to see three more cycles come round, so he must have been satisfied in his old age that the Sun has its own rhythm. (Actually, several other possible cycles have since been identified.)

Sunspot data, although sporadic, was available from the seventeenth century onwards but data for magnetic variation only from the 1830s. Wolf had difficulty in tying the two together. He had two arguments to show that there was a connection between changing features on the Sun and changes in the magnetic field on Earth. The first argument was rather weak. It concerned a sudden short-lived, intensive flash of light seen on the Sun by Carrington on September 1, 1859, followed by a magnetic storm on Earth and some days later by spectacular auroras as far south as Hawaii and Cuba. Wolf took this to be a direct proof of a possible relationship between the Sun and terrestrial magnetism, but he did admit that *post hoc* is not necessarily *propter hoc* and more such events would be necessary for a real connection to be established.

Wolf's second argument was that a prediction he made had come true. This he called his indirect proof. Based on his 'relative sunspot numbers' Wolf calculated ahead of time what the magnetic variation would be in 1859. Accurate observations verified his forecast. Although making such predictions and testing them

THE ENIGMA OF SUNSPOTS

is the best way to make sure that one is on the right track, this did not mean that his theory was necessarily correct and that a long term correlation between sunspot numbers and the changes in magnetic variation really existed.

Wolf's guess was correct, but a guess it remained because he still had no answer to the question of why the sunspot cycles correspond to the cycles of geomagnetic field variation. By way of a slight philosophical digression here, we need to review the several possibilities why the courses of two natural processes could resemble or, track each other. It is but an extension of Sabine's ideas as expressed above.

Firstly, the two may have nothing to do with each other, their correlation being pure coincidence. The second possibility is that one process influences the other. Thirdly, there could be an extraneous factor influencing both processes.

Here are some examples. In a textbook of elementary statistics the following exercise is set: 'Explain why it would not be surprising to find a fairly high correlation between the density of traffic on Wall Street and the height of tide in Maine if observations were taken every hour from 6.00 am to 10.00 pm and high tide occurred at 7.00 am.' (Americans keep earlier hours than the British.) This exercise could be extended to several days for a place with fairly regular twice daily tides, but it would obviously break down after five consecutive weekdays. In general, a connection between rush hour traffic and the tides in the vicinity of a sea-port would be understandable, but not at such distant locations. As this example shows, pure coincidence has the tendency to break down after a while. If something we classed as coincidence persists, we soon start looking for an explanation to show that it was not really a coincidence.

A lighthearted tribute for the eightieth birthday of an outstanding solar-terrestrial physicist, who had published more than 450 articles, supplies yet another example. In the article entitled: 'Correlation of sunspot numbers with the quantity of S. Chapman's publications,' the author shows that there is good correlation between the two, but with a phase difference of two years.[7] The time lag of two years between the submission of a

Details of sunspots and granulation.

Above: *NASA*

Below: *U.S. National Solar Observatory, Sacramento Peak)*

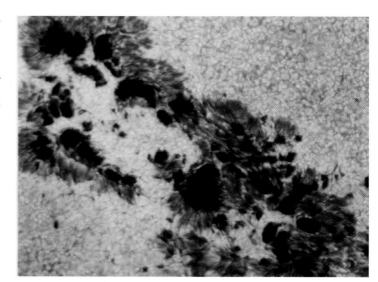

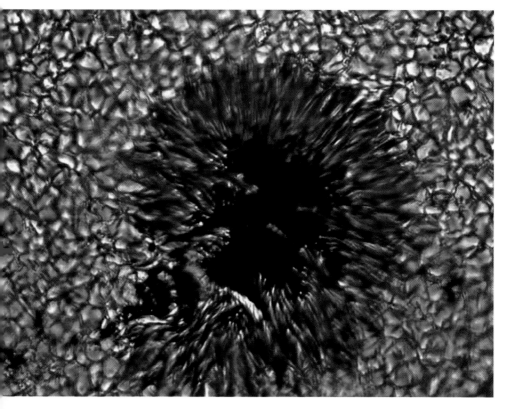

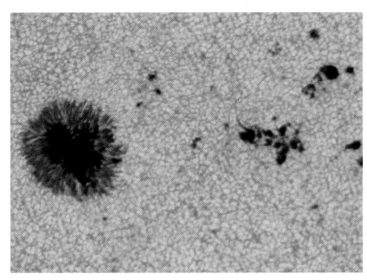

Above:
*Detail of sunspots
and granulation.
(U.S. National Solar
Observatory,
Sacramento Peak)*

Below:
*A spiral sunspot.
(U.S. National Solar
Observatory,
Sacramento Peak)*

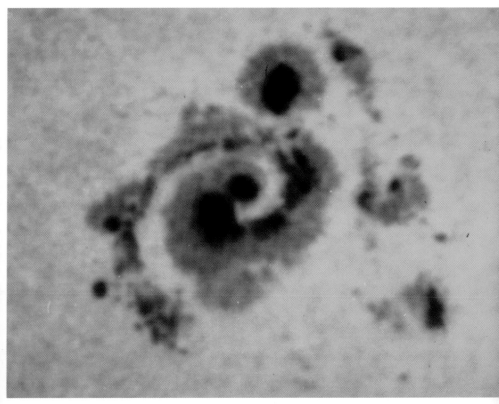

paper to a journal and its publication neatly explains the phase difference. Another example, often cited in elementary textbooks on statistics, is the old chestnut about the reading ability of children matching the ascending curve of their shoe sizes. It does not require deep analysis to see why this is so: the fact that the children are getting older accounts for both.

My last example comes from real sunspot studies. When plotted on a graph, the water level of Lake Victoria in Africa matched solar activity well enough between 1896 and 1922 but showed no relationship to it after that date. Was the relationship a coincidence, or did some fundamental changes occur after 1922?

Because Wolf and his colleagues could not account for any physical mechanism between sunspots and changes in the direction of the compass needle, all the above three possibilities remained open. Was the similarity between the sunspot cycle and the magnetic variation cycle just a spurious association? (Like that between the tides in Maine in a particular week and rush hour traffic in New York). If this were the case, then we would expect the similarity of the cycles to cease after a while. This did not happen.

There could be an external influence acting in tandem on both. (As a child grows older his feet get bigger and his reading ability improves.) For a while Wolf entertained the idea that both the spots and our magnetic field are due to planetary feedback, especially to the influence of Jupiter. This was not an irrational idea, because it cannot be ruled out that an occasional alignment of the planets can have some influence both on solar and terrestrial activity.

Maybe changes in terrestrial magnetism could bring about changes on the Sun? Not impossible, but it is difficult to believe in any major influence if we consider the difference in size between the two bodies. The Earth influencing the Sun is difficult to believe for psychological reasons as well, because all through history until the most recent times, human beings knew that however much they prayed or worked special magic, they were really powerless in the face of natural forces. The very last possibility is that the sunspots influence the magnetic field. Wolf

THE ENIGMA OF SUNSPOTS

did not find it impossible that the Earth's magnetic field is actually produced by the Sun and its changing spots. Some others however, discarded any kind of causal connection between dark areas on the Sun and the magnetic field here on Earth because it sounded very much like an unwelcome leaning towards astrology. The answer to the question of what makes the two cycles coincide had to wait for another clue to be found, for yet another discovery to be made.

128   *THE ENIGMA OF SUNSPOTS*

∾ 13 ⊌

A Sudden Flare of Light

Although in the middle of the nineteenth century the discovery of the sunspot cycles introduced some measure of regularity into the behaviour displayed by the Sun, still no explanation was forthcoming of how and why, if at all, the Sun's activity had an effect on the geomagnetic field. Not even when Richard Carrington's above-mentioned discovery of a sudden event on the Sun (see p.122) was followed by magnetic turmoil, could anyone describe the physical mechanism of how solar events can cause disturbances on the Earth. In fact, this was not elucidated until the middle of the twentieth century.

Carrington was another of the grand amateurs. He was born in 1826 into a family of brewers just like Hevelius and for a while he could devote his time, money and energy to astronomy. It was Carrington who so kindly went to Petworth House in Sussex at Wolf's bidding and copied Harriot's notes. He built himself an observatory in Redhill in Surrey and equipped it with telescopes made by the best instrument makers in England. He resolved to be absolutely methodical in his research and faithfully kept to this resolution, taking over 5000 regular observations. Carrington could only watch the sky for less than one solar cycle, because when his father died he had to take over the running of the brewery and had no time left for astronomy. Despite this, his observations and his calculations brought excellent results.

As soon as he could, Carrington sold the brewery business and started building another observatory near Churt, on a steep hill, the Devil's Jump, also in Surrey. But this new endeavour was to be sadly cut short. Carrington had contracted an unfortunate marriage. When his wife was found dead from an overdose of medicaments, although Carrington was not accused of her

The aurora is intimately linked to sunspot activity. (From H. Falck-Ytter, Aurora)

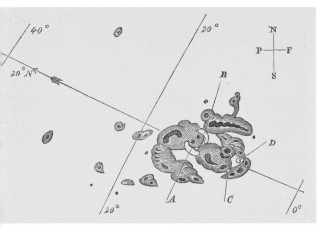

murder, the coroner nevertheless reprimanded him for not taking good care of her. Carrington was by then already a sick man but now his spirits broke as well and he died soon afterwards, in 1875.

The solar event that Carrington discovered and Wolf referred to, was a solar flare. As its name suggests, a flare is a short-lived, sudden, intense brightening of a small region. On September 1, 1859 Carrington, following his daily custom, was watching the Sun when he suddenly saw two flares erupt within the area of a large sunspot group. He jumped up and ran outside to find an independent witness to this event but by the time he returned, barely a minute later, the apparition was already fading away. Carrington was a reliable observer so he would have been believed anyway, but luckily another amateur astronomer, Richard Hodgson, could attest that he saw the same flare at his home in Highgate. Actually Stephen Gray, owner of a not very successful dyeing business in Canterbury had already seen such a flash in 1705 but no one took much notice of it. Gray, who is better known for his electrical experiments, was also a sunspot watcher and one of Flamsteed's loyal correspondents. As

Above:
The flares Carrington saw were in the position of A & B. By the time of his return they were faint and had moved to position C & D.

Below:
Nineteenth century sunspot drawings by Louis Trouvelot

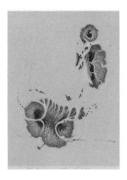

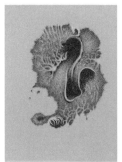

THE ENIGMA OF SUNSPOTS

such he was no favourite of Newton's who may have prevented the publication of some of his communications to the Royal Society.

Flares are relatively frequent but short-lived and rarely bright enough to be distinguished against the brilliant disc of the Sun. That is why it took so long to bring them to public attention. Following the appearance of the flare that Carrington saw in September 1859, the Earth experienced a massive magnetic disturbance. Previous such incidents, prior to regular observations of terrestrial magnetism and the widespread use of electric appliances, were not noticed. In 1859 telegraphy, operational by now, was severely disrupted. Auroras were seen all over England.

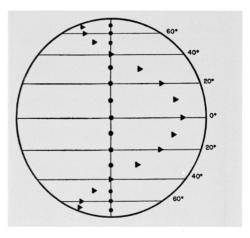

After 30 days spots that, hypothetically started on the midline (circles) would approximately be in the positions indicated by the triangles. This is because those near the equator move faster. (from H.T Stetson 1947)

Did the flare cause the magnetic disturbance? Skeptics remained skeptical of any causal relationship and remained convinced that it was just a coincidence that the unusual magnetic event happened so soon after the flare was seen. Carrington himself expressed doubts of any connection between the two. One swallow was not enough for him. Then between 1873 and 1892 Edward Maunder, astronomer at the observatory at Greenwich, noticed three magnetic storms coinciding with remarkable sunspot activity and concluded that a connection between the two events was more than likely. Some people remembered the old discarded idea of solar exhalations and were already thinking in terms of particles in space being the culprits. It took another hundred years to find the particles carrying solar magnetic fields, but the trail was getting warmer all the time.

Seeing a solar flare was a bit of luck for Carrington but his careful studies led him to other discoveries as well. Scheiner had already noted that the spots do not travel with equal speed across the disc of the Sun, those closer to the equator move faster than the ones closer to the poles. Astronomers made valiant efforts over the years to determine how long it takes for the Sun to turn

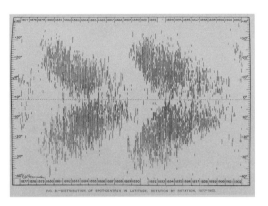

FIG. 8.—DISTRIBUTION OF SPOT-CENTRES IN LATITUDE, ROTATION BY ROTATION, 1877-1902.

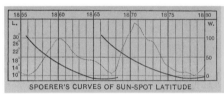

SPOERER'S CURVES OF SUN-SPOT LATITUDE

Above:
*Edward Maunder's
butterfly diagram*

Below: *Spoerer's
curves of sunspot
latitude*

around, but the values they came up with varied according to the position of the spots they studied. As a result, there was general disagreement about the period of rotation. In the middle of the eighteenth century Jean Bernoulli the Younger, member of the famous Swiss family of scientists was planning to write a guidebook for tourists wishing to visit European observatories and went on a fact-finding tour. He reported that several scientists tried to calculate the period of the Sun's rotation from three observations taken of the same spot at different locations. These calculations, however satisfying they may have been from a geometrical point of view to the frustrated astronomers, added little information about the real world.

Carrington on the other hand based his calculations on precise measurements of several spots at several different latitudes and found that the equator turns in about 25 days, but closer to the poles the Sun's surface is slower moving and takes much longer to complete a revolution. The Sun turns faster in the middle and slower at the poles. Modern methods using Doppler shifts to measure the Sun's rotation put the polar rotation at about 35 days.

The Italian astronomer Angelo Secchi, who spent much of his life studying the Sun, pointed out that the rapid changes that sunspots are prone to, make it certain that the photosphere cannot be solid. He calculated how a ball made of an 'elastic fluid' would rotate and came up with figures similar to those of Carrington's. In 1864 Secchi announced that the Sun is gaseous. We now believe that Secchi was right and the Sun is indeed a ball of gas. But the gas that makes up the Sun is no ordinary gas. Being extraordinarily hot and dense, it is unlike any gas encountered on Earth under natural circumstances. Made of gas, the Sun does

THE ENIGMA OF SUNSPOTS

not rotate like a solid ball, some parts move slower, some faster. This has important consequences: the different rotational speeds both inside the Sun's body and on its surface result in strong and changeable magnetic fields. In addition, sunspot studies show that the rotational speed of the Sun is not constant but varies over time.

Carrington noticed yet another peculiarity. Near the maximum of the solar cycle, that is when sunspots are many, new spots appear not too far from the Sun's equator.

The photoheliograph.

There is a slight overlap when an old cycle ends and a new cycle begins and at that stage spots tend to appear at higher latitudes, nearer the Sun's poles. Since Carrington had to turn to brewery before he could complete his observations for a full solar cycle, this phenomenon is now not known as Carrington's Law but as Spoerer's Law after the mathematics teacher turned astronomer who made a proper study of it. Solar scientists who try to predict the behaviour of the solar cycles find tracking the latitudes where spots appear very useful, because spots appearing further from the equator herald the beginning of a new cycle.

It was Maunder, who found an ingenious, and rather beautiful way of demonstrating Spoerer's Law. This is the 'Maunder Butterfly Diagram.' Comparison of this striking illustration with other diagrams of the same phenomenon demonstrates what a difference a well chosen representation can make to our understanding.

Maunder used the photoheliograph, an instrument devised and constructed by Warren de la Rue for taking photographic pictures of the Sun.* The invention of photography brought great advances in astronomy and especially in sunspot studies. A photograph is more objective than a personal observation. It can be studied at leisure and several exposures can be taken at intervals. Also, using light from those parts of the spectrum that

* It was paid for by the Royal Society at the instigation of Sir John Herschel, astronomer son of Sir William. John Herschel catalogued the stars of the Southern Hemisphere and studied sunspots at the Cape in South Africa.

the retina can hardly, or not at all register, a photograph can reveal features not seen by the eye. Jules Janssen of the observatory at Meudon in France was convinced that photography would serve as a new method of making discoveries. To prove his point he took photographs of two large spots in June 1885 and showed that the structure of the penumbra and the area connecting the spots was identical to that of the rest of the photosphere. Although by then it was unlikely that spots could be clouds, or accretions of material, or uncovered mountain peaks, or chasms, this was the first actual proof that their penumbra is much like the rest of the Sun's surface.

Actually the first photograph of the Sun, a daguerreotype, was made in 1845 by two French physicists Fizeau and Foucault. Then just by chance, at the Great Exhibition in 1851, some photographs of the Moon were displayed close to the envelope-folding machine invented by Warren De La Rue. As he walked past, the idea of photographing celestial objects suddenly excited this rich stationery manufacturer. De la Rue employed the more advanced collodion wet-plate process and solved the problem of producing the extremely short exposures that were needed. From 1858 onwards he used his photoheliograph regularly at the Kew Observatory. In 1861 he succeeded in taking stereoscopic pictures of the Sun and proved that Wilson was right: the spots showed up as indentations.

The observatory at Greenwich where Maunder worked was originally founded for a single strictly practical purpose: to find the exact positions of the Sun, the Moon and the stars. Without this knowledge navigation at sea could at best be described as uncertain and often dangerous. Later, with the arrival of ships made of iron, a magnetic department was installed, also in aid of navigation. Regular study of sunspots started only in 1873 when the photoheliograph was transferred there from the Kew Observatory and

THE ENIGMA OF SUNSPOTS

Edward Maunder was appointed as a 'photographic and spectroscopic assistant.' This addition to the observatory's research programme came fully twenty years after the discovery of the relationship between the Sun and terrestrial magnetism, but government institutions are known to move slowly. The main reason for setting up the new department was not inspired by any wish to improve our knowledge of the world but by an announcement of C. Meldrum, director of the observatory in Mauritius that a connection existed between the sunspot cycle and cyclones in the Indian ocean.

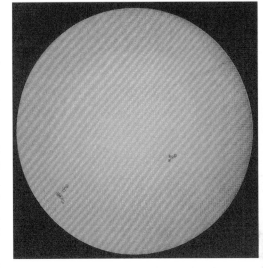

The very first daguerrotype image of sunspots.

It was hoped that the study of the Sun would lead to the correct forecasting of cyclones with attendant great economic advantages. A second solar observatory was established in 1879 in South Kensington. This was the outcome of an investigation by the Royal Commission on Scientific Instruction and the Advancement of Science, set up because science in Britain was believed to be lagging behind that of its continental neighbours. The Solar Physics Observatory in Kensington, or the 'sun-spottery' as it became known, started out at first as just a collection of wooden and canvas huts. Sir Norman Lockyer, originally a civil servant and at first only an amateur astronomer, gained such a reputation that he was appointed as director. The Kensington sun-spottery concentrated on solar chemistry and later on the connection of sunspots with the weather.*

At Greenwich Edward Maunder took daily photographs of the Sun whenever he could. Greenwich is not exactly famous for its sunny climate and so Maunder arranged to have photographs taken in India and in Mauritius as well as at Greenwich. In spite of his rather unflattering description of the work of an astronomer as the combination of engineering and accountancy, Maunder was as enthusiastic an observer as can be.

* Lockyer resigned in 1911 and set up an observatory near Sidmouth in Devon. A centre of research until the 1980s, this is now an educational and tourist attraction run by volunteers.

De La Rue's July 1860 eclipse party in Spain with heliograph.

The observatory at Greenwich employed a posse of uneducated young boys at a very low salary as 'computers', to carry out arithmetical operations. A professor of astronomy at Oxford explained: 'The method of using the few years between 14 and 17 or 18 in the life of a boy who has no immediate prospects seems the least wasteful.'[1] In other words, after a few years the boys could simply be thrown out and replaced with new, young recruits and in this way wages never had to be raised.

In 1890 a decision was made to add a number of graduate women to this group of computers. No one thought it strange to employ university educated women on a par with uneducated lads.

Initially four women were offered a post. After lengthy consideration one of them, Agnes Clerke, the historian of astronomy, refused the offer. Her reason was not that she found it demeaning, but she considered Greenwich unsafe for a woman at night,

THE ENIGMA OF SUNSPOTS

and in any case, by then she was already making a good living by her writing.

In 1891, at the resignation of one of the originally employed 'Lady Computers,' Annie Dill Scott Russell was appointed. Born in Ireland in 1868, she had graduated 'Senior Optime' in mathematics, the highest degree a woman could then achieve at Cambridge. On her appointment to the Greenwich Observatory she had to take a cut in salary from her previous job as a schoolmistress, which paid £80 per year plus board, to a miserable £48 per year.

Edward Maunder.

Low status and low salary were not the reasons why Annie Russell only stayed in the job for four years. In 1895 she married Maunder and according to Civil Service rules she had to resign. She returned to work as an unpaid volunteer in 1916 during the First World War, when lack of manpower made her services badly needed. The men in charge of the observatory could be assured that she was still fully competent because even after her marriage she never stopped her astronomical work.

In the Maunders we meet yet another example of an astronomical couple. Annie Maunder measured and computed the positions of sunspots from the photographs taken by her husband. She also took part in solar eclipse expeditions where she took her own photographs and conducted her own experiments. In 1898 in India 'Mrs Maunder solved the problem of photographing the faint coronal extensions,' the hardly visible long rays that shoot out from the rim of the Sun. According to a report of the 1905 eclipse, when the couple went to Mauritius, Maunder caught a bug and was so sick that 'the developing of eclipse photographs has had to be left almost entirely in the capable hands of Mrs Maunder and Mr A Walter.'[2] As it turned out, the photographs taken by Annie Maunder were by far the best ones produced by the entire expedition.

In 1908 the Maunders published *The Heavens and their Story* jointly but Maunder acknowledged in the preface that the book was almost wholly the work of his wife. Annie Maunder makes an interesting suggestion in the book. It seems that more sunspots go into oblivion when they are behind the Sun than when they are on the visible side. Could it be after all that the Earth does have an influence on the spots? At the moment it seems not. Mrs Maunder's figures were for the years 1889–1901; a new calculation made for 1944–75 does not show any such influence.

In addition to the butterfly diagram, Edward Maunder's name is now firmly attached to a solar event that took place in the second half of the seventeenth century: the Maunder Minimum. Before discussing it we shall have to turn to the momentous discoveries that were made even before Maunder's appointment at Greenwich. These discoveries were not strictly speaking in astronomy, rather in physics and in chemistry, but they impacted on solar studies. They opened up new fields and yet again changed the direction of research.

∾ 14 ↢

The Magnetic Sun

Light travels across space. That is why we can see the Sun many millions of miles away. Radiant heat also travels through space. That is why Joseph Henry was able to measure the temperature difference between two areas on the Sun's surface. Would it be possible to get information, or take measurements, of any other chemical or physical properties of the Sun? Properties other than heat and light.

It is possible. Surprising as it may seem, astronomers can now tell what chemical elements the Sun and the stars are made of, without having to visit them personally. And in 1908 the American astronomer, George Ellery Hale, was able to affirm the existence of a magnetic field on the Sun by observing something here on Earth. Sitting in his observatory Hale detected the effect of the Sun's magnetic field on its atoms.

There is a simple test for finding out if an object is magnetic or not. If it picks up pins or nails placed close to it, then we know that it is magnetic. Obviously the same cannot be done with the Sun. In order to see if the Sun is magnetic or not, a roundabout way had to be found. This roundabout way starts with a polished piece of glass or rock crystal hung on the window that produces a lovely rainbow effect when the Sun shines on it. When we behold a rainbow in the sky the raindrops have performed the same function for us. Newton let the light of the Sun through a chink in a shuttered room and used a glass prism to split it into colours. (To prove his point he collected the colours together into white light again with another prism.) We call the artificial rainbow thus created the spectrum. In the spectrum the different colours represent light of different wavelengths.

A glowing solid or liquid body emits so many colours that

they all run together and form a continuous spectrum. But in 1859 Gustav Kirchhoff and Robert Bunsen (inventor of the Bunsen burner), professors at the prestigious university in Heidelberg, Germany, realized that when individual chemical elements, for instance sulphur or calcium, are heated to such a high temperature that they are incandescent, then the spectrum of the light they emit is not continuous but consists of separate, discrete lines. The position of these lines (corresponding to their wavelength) is characteristic of the particular element. Conversely it is also true, that the atoms of an element in a gaseous state will absorb light of exactly the same wavelengths that they can emit. These show up as dark 'absorption lines' in the spectrum. In 1802 the medical man turned chemist W.H. Wollaston and some ten years later the optician Joseph von Fraunhofer saw such dark lines in the spectrum of the Sun. Fraunhofer even studied the lines closely, listed them and gave them their names, but did not recognize their practical use. Kirchhoff and Bunsen noticed that the lines emitted by the substances they tested coincided with some of the dark Fraunhofer lines. As Kirchhoff put it: 'The dark lines of the solar spectrum ... exist in consequence of the presence, in the incandescent atmosphere of the Sun, of those substances which in the spectrum of a flame produce bright lines at the same place.'[1] With Kirchhoff's insight astrophysics was born, because by looking at the spectrum of the Sun, or that of a star, we can now tell what they are made of.

The discovery of spectroscopy is an object lesson that in the scientific endeavour trying to foretell the future is always fraught with danger. Only a few years earlier the positivist philosopher Auguste Comte declared that we shall never be able to find out anything about the true nature of the Sun or that of the stars. In Harriet Martineau's translation: 'We can never learn their internal constitution, nor, in regard to some of them, how heat is absorbed by their atmosphere.'[2] Never say never, especially in science.

As a philosophical fallout, spectroscopy gave the opportunity for the first time ever, to really prove, that Aristotle was wrong in surmising that the Sun is made of a completely different material

THE ENIGMA OF SUNSPOTS

from that found on Earth. The discovery of sunspots made it likely that he was wrong, but definitive proof was lacking before the invention of spectroscopy. True, by the nineteenth century such a comprehensive refutation was not needed any more. Aristotelian dogma was dead, and well and truly buried. (Scientists do not subscribe to the view that a long held belief must necessarily be true.) By then scientific opinion had swung to the opposite extreme, helped by the theory put forward by Immanuel Kant in 1755 and forty years later by the French scientist Pierre Laplace. If the whole solar system, including the Sun, developed from a rotating gaseous nebula, it was imperative that all of it must consist of the same material. This view was so popular that when Sir Norman Lockyer, director of the Solar Observatory at South Kensington (and founder of the journal *Nature* which he then edited for the next fifty years), studied the lines in the solar spectrum and reported finding new elements there, hitherto unknown on Earth, scientists would not believe him. Sure enough, Lockyer was mistaken and all his 'new' elements were satisfactorily identified with well-known terrestrial elements. All, except one. This element, helium, was not yet known on Earth. It took a long time, nearly thirty years after Lockyer's first announcement, for helium to be detected down here. Luckily Lockyer was then still alive. (The element was named helium from the Greek word for the sun: *helios*).

As things stand at the moment, even if a substance that does not exist on Earth were to be found in the Sun or in any of the stars, it would not induce scientists to change their minds and subscribe to Aristotle's theory. We have come to believe that the physics and chemistry of stellar material is not any different from the physics and chemistry of terrestrial material. So how can this be reconciled with the fact that in the twentieth century science seems to have made a somersault once again and postulated that some of the physical and chemical processes in the Sun and in the stars do differ from those naturally occurring here on Earth? Does that mean a return, albeit in a new version, to the ancient perception that natural laws are different here than in the heavens?

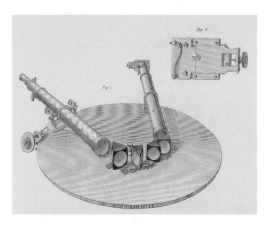

The spectroscope Gustav Kirchhoff used to produce a solar spectrum.

It does not. For example, the Sun and the other stars produce their energy by nuclear fusion, a process that does not naturally occur here, but that the successful production of the hydrogen bomb proved is possible to replicate. Also, the spectrum of the Sun is continuous, like that of a solid or a liquid, not like that of a gas which consists of separate lines. How come then that we believe the Sun to be a ball of gas? The answer is that the gas of the Sun is like no gas naturally found on Earth, it is so hot and so condensed that it can emit a continuous spectrum of light. But again, if we successfully produce conditions similar to those existing in the Sun, we expect, and get, the same outcome. We think of the universe as one unit and it would take fundamentally new discoveries buttressed by a fundamentally new theory to shake us from this conviction. This is not impossible but unlikely in the near future.

In the middle of the nineteenth century arguments and controversies flourished just as much as before, but the new methods of spectroscopy and photography could now help in deciding between conflicting theories. The French astronomer Hervé Faye believed that the flow of material at the location of spots was upwards; by contrast the astronomers at the Kew observatory were convinced that the flow of material had a downward trend. Spectroscopic measurements showed that the latter was correct. Janssen declared that when the Sun is not very active spots appear and disappear faster than when the Sun is active, and produced photographs to quell the critics who accused him of lax observation. James Nasmyth, inventor of the steam hammer, initiated a debate about 'willow-leaves' he claimed to have seen over the whole surface of the Sun. They could be giant strange organisms, speculated Sir John Herschel. When Sir Norman Lockyer observed the pattern forming only in the penumbra of spots, this particular idea faded away. Secchi favoured the theory

THE ENIGMA OF SUNSPOTS

THE EARTH
TO THE SAME SCALE

that spots were accumulations of condensation products from clouds of gas called prominences.

James Nasmyth's painting.

A turning point came when using spectroscopic methods magnetic fields were detected in the sunspot areas. What has spectroscopy to do with magnetism? How could looking at the light coming from the Sun help us to find out if there is any magnetic activity on it? Unlikely as it may seem, this is exactly the case. The missing clue was supplied by a discovery for which the Dutch physicist Pieter Zeeman received the Nobel Prize in 1902. He found that if the light emitting or absorbing material such as for instance, a sodium flame, is in a strong magnetic field, this will interfere with the spectrum and widen or split the lines in it. Michael Faraday had suspected this effect and tried to find it in the laboratory but failed. On the other hand several people had previously reported seeing the widening and splitting of the spectral lines coming from a spot area without realizing what that meant. Lockyer, who was the first to study sunspot spectra (and after Janssen the first to photograph prominences without an eclipse, proving without a doubt that they belonged to the Sun and not to the Moon), put down the widening of lines to heat pulses in the Sun. His theory was that these pulses caused temperature changes in the spots; they widened the spectral lines, and at the same time were responsible for changes in the weather.

It took twelve years from the announcement of the Zeeman effect but at last the magnetism of the Sun, or even that of a star much further away, could be revealed. Why did it take so long when all one had to do was to look at the spectrum and see if any lines had been split? Because it is easy to say, not so easy to do. It needed the ingenuity of someone to devise a new instrument and the finesse to obtain funding for its construction.

While Warren de la Rue was not short of money to at least partially fund his own projects, Hale was an excellent fundraiser and an organizing genius. Thanks to him the International Union for Cooperation in Solar Research came into being. The sponsorships he raised helped build the Yerkes Observatory in Chicago and the observatories in California on Mount Wilson and on Mount Palomar.

In previous ages it was the quality of the instruments that would determine what sort of observations astronomers could perform. Hale on the other hand, had his instruments built specially for the problems he set for himself to investigate. For the observations he wanted to make, just like Hevelius before him, he had huge telescopes constructed. Unlike his predecessor's instruments however, those built by Hale are noted not for their length but for the size of their optical components and are eminently usable — at least they were usable until the air pollution of Southern California caught up with them. Hale pioneered tower telescopes in order to reduce the disturbance caused by movement of air near the ground.

With his spectroheliograph, a special adaptation of the photoheliograph, Hale was able to prove the hunch he had that sunspots have a strong magnetic field. The story goes that the idea for the instrument came to him while travelling on a tram — he was looking at a picket fence and the garden beyond as it appeared bit by bit through the gaps. With the introduction of an additional slit and scanning the surface of the Sun, the spectroheliograph could take photographs in monochromatic light, that is, using one wavelength only.

In 1908 Hale announced that the effect Zeeman discovered was plainly visible in the light coming from the sunspot area. That area was magnetic. Then in his laboratory Hale compared the sunspot spectra with that of luminous gases in magnetic fields. He found that he had to use extremely strong magnetic fields if he wanted to get the same degree of line splitting in his lab experiments as the light coming from the sunspots he studied produced. This meant that the sunspots also had very strong magnetic fields.

George Ellery Hale's photographs of bipolar spots.

The president of the Carnegie Institution wrote: 'This is surely the greatest advance that has been made since Galileo's discovery of those blemishes on the Sun'[3]. Indeed, Hale's discovery opened up new avenues in solar research. Investigations at once started to focus on the magnetic activities displayed by the Sun and sunspots took their place once again at the cutting edge of research.

Hale discovered something else as well. Spots often come in pairs, one leading the way and the other following. Hale found that these twins have opposite magnetic qualities: if one is 'north,' then the other is 'south' (or + and -). Furthermore, this magnetic polarity changes every eleven years, so in twenty-two years the original polarity is established again: yet another cyclic change on the Sun, and exactly twice as long as the sunspot cycle. We would probably be calling this the solar cycle if the concept of the sunspot cycle had not been so firmly established by then.

Time to return to the second effect to which Maunder's name is attached. This concerns a strange period in sunspot history. As we have seen, in the second half of the seventeenth century not many spots were seen and the few that were seen, were rather small. This was widely known and accepted but carried little significance. In the absence of any scientific interest in the extended minimum, a theological argument was put forward against Herschel's inhabited Sun. In the eight years between 1676 and 1684 no sunspots were reported. If there were creatures living on the Sun, their windows to the sky would be the sunspots and surely God would not exclude any of his rational creations for such a long time from the 'contemplation of the starry firmament.'

As time went by, the extended minimum faded from consciousness. Gustav Spoerer, who made his own historical researches, pointed out the anomaly in 1887 and again in 1889. He even suggested that Hevelius gave up his sunspot observations because there were so few sunspots to be seen. No one seemed to be interested and Spoerer's announcement of yet another sunspot minimum between 1480 and 1520 possibly only weakened his case. When Edward Maunder tried to draw attention to Spoerer's paper and to the strange period in the seventeenth century he was not much more successful either. In 1894 Maunder published an article about the extended minimum in the August issue of the popular science magazine, *Knowledge*.[4] Following Maunder's article Agnes Clerke pointed out in a letter to the editor that 'in England, during the whole of the seventeenth century, not an auroral glimmer was chronicled.'[5] Since by then a connection between magnetic disturbances and auroras was reasonably well established, she concluded that the prolonged sunspot minimum was attended by a profound magnetic calm.

Maunder did not give up. After a lapse of some thirty years

he made another stab at drawing attention to the extended sunspot minimum but his note published in 1922 in the *Journal of the British Astronomical Society* again failed to excite the imagination of the scientific community.[6] The scientist A.E. Douglass whose interest in the Sun and in climate led him to develop tree ring dating just a few years earlier, was one of the few men who responded to Maunder's article. Tree rings are laid down annually and their width and chemical composition can serve as a record of the weather. Working backwards in time from trees of known age to older ones and comparing the patterns of tree rings, Douglass introduced a method to date wooden objects. Carrying the dating back for 500 years he found a cycle of 11.3 years and was persuaded that he managed to correlate the tree ring patterns with the sunspot cycle. On reading Maunder's article he suddenly recalled that the period between 1670 and 1727 had given him trouble because the curve of the cycles seemed to flatten out so much that no cycles could be detected at all. Maunder's contention of the lack of sunspots in roughly the same period confirmed for both that the minimum existed.*

A full hundred and fifty years had to go by before the penny finally dropped, that such an event might have serious significance. If for half a century there were hardly any sunspots, that is, hardly any solar magnetic activity, then the Sun's behaviour was completely different from its present behaviour. And, if the Sun behaved in such an unusual way in the seventeenth century, then it might do it again or, it could play some other trick on us. What is certain, if what we now call the 'Maunder Minimum' really took place, is that the Sun is a much more changeable star than it was held to be.

Did the Maunder Minimum really exist? In 1976 John Eddy, a scientist working in Boulder, Colorado, carefully listed the arguments both for and against its existence and came to the conclusion that it must have happened.[7] Warnings of greenhouse gases were just beginning to be issued and this may have been one of the reasons why Eddy had more success than Maunder in drawing attention to the extended minimum and its

* Later studies have shown evidence for weak tree ring cycles. This is not a refutation of the Maunder Minimum, just that non-existent tree ring cycles cannot be used as proof of its existence.

possible influence on the climate. Another reason was that Eddy was able to support his arguments by measurements not yet available to Spoerer and Maunder: the abundance of the carbon-14 isotope in tree rings. Here we are in need of some explanation. The Earth is constantly bombarded by extremely energetic particles known as the cosmic rays. Rays is a misnomer because cosmic rays are not rays at all. They consist of electrons and ions (atomic nuclei stripped of electrons) and travel very fast. Some originate in our galaxy but most of them come from outside the solar system, probably from exploded stars. Cosmic rays are capable of entering and modifying atoms and they produce isotopes such as for instance the radioactive carbon-14. Trees take in carbon dioxide from the atmosphere and deposit the carbon in their annual rings. The amount of carbon-14 compared to ordinary carbon in the rings depends in large measure on the amount of cosmic rays reaching our atmosphere. When there are few cosmic rays, not much carbon-14 is produced and deposited, when there are many cosmic rays reaching us then more carbon-14 is produced and deposited in the tree rings. The rings can be dated and from the amount of carbon-14 they contain we can find out if cosmic rays were abundant during a particular year, or not.

When there are many spots, faculae, flares and other features on the Sun, in other words when the Sun is active, it occasionally emits large masses of energetic particles. They can disturb the normal pattern of the magnetic field surrounding the Earth, cause magnetic storms and corresponding trouble. They also drive away the cosmic rays and prevent them from reaching us. Fewer cosmic rays in turn mean less carbon-14 produced.

What can we expect when the Sun is not very active? Few spots, little emission, but lots of cosmic rays getting through, and an abundance of carbon-14 in the tree rings. This is exactly what was found for the second half of the seventeenth century and as a consequence the existence of the Maunder Minimum is now generally accepted. Tree ring data did help, but not the way Douglass imagined it.

THE ENIGMA OF SUNSPOTS

As a nice touch, Eddy pointed out that the Maunder Minimum coincided with the reign of Louis XIV, the 'Sun King.' Let's finish this chapter with Eddy's insightful words:

> The reality of the Maunder Minimum and its implications of basic solar change may be but one more defeat in our long and losing battle to keep the Sun perfect, or, if not perfect, constant, and if not constant, regular. Why we think the Sun should be any of these when other stars are not is more a question for social than for physical science.[8]

∼ 15 ∽

A Glance at the Sun

At the beginning of the twentieth century it was known that the Sun was a rotating ball of gas and that sunspots were indented cooler regions with magnetic properties. The chemical elements present were identified but their relative abundance was not known. No firm ideas of the structure of the Sun existed. Most importantly, it was still a 'grand mystery' how the Sun (and by implication any star) produces its immense amount of energy.

By the end of the century our understanding had radically increased. What brought about the change? Fundamental theoretical advances based on experimental results and impacting on solar studies were: Einstein's theory of special relativity postulating the equivalence of matter and energy (the famous $E=mc^2$ equation), quantum theory and nuclear physics. Increasingly effective use of applied mathematics and mathematical modelling is another important theoretical factor. On a more practical level, the introduction of ever more sophisticated instruments and advances in rocket engineering produced revolutionary observational techniques, necessary to the building of new theories and hypotheses.

A detailed account of twentieth century solar science is beyond our scope. But a simple overview of contemporary ideas and unresolved problems concerning the Sun and its spots will help explain some of the phenomena we have encountered. It will also make us look at the Sun with renewed respect and amazement. In his book *Unweaving the Rainbow*, Richard Dawkins argued that scientific enquiry does not deprive us from experiencing the wonder of the natural world. On the contrary, the heart leaps up at the sight of a rainbow all the more if one has an understanding of how it is created. Dawkins' argument is directed against those who believe with D.H. Lawrence that 'knowledge has killed the Sun, making it a ball of gas with spots.' I find myself in agreement with Dawkins on this point. Is not the image of the Earth making a complete revolution in space each day with all of us riding on it more exciting than the Sun sitting in a cave at night? Could stories about chariots in the sky be more awe-inspiring than the explanation of how the Sun is able to produce such an enormous amount of heat and light for millions and millions of years? And surely, I was only one of many whose elation was enhanced while watching the Moon slowly creep over the disc of the Sun during the 1999 total eclipse, by the knowledge of what was happening, by our ability to gain that knowledge, and by the accuracy of the forecast.

Early in the twentieth century the study of spectral lines established that the Sun consists mostly of hydrogen. The next abundant element is helium. Heavier elements are also present but in much smaller amounts. In the middle of the century the problem of the Sun's energy production was satisfactorily solved — although future generations may think otherwise. We are reasonably certain that the Sun produces its energy by nuclear fusion, a complex nuclear reaction that takes place deep inside the Sun (making the bumper stickers some cars used to parade in the 1980s that had a smiling Sun and the caption: 'Nuclear energy? No, thank you' somewhat ridiculous). Energy liberated in the core of the Sun by the fusion of four hydrogen atoms into one helium atom is carried outwards, at first by radiation. In the next surrounding layer energy is transmitted by moving matter (convection). Movement of this turbulent mass is a difficult physical and mathematical problem and is as yet not entirely solved. The visible surface of the Sun, called the photosphere, is much cooler than the core, though still very hot. This layer radiates most of the energy into space. The photosphere is surrounded by another layer of gas, the chromosphere; and the Sun's outermost envelope is the corona. We can not normally see these outer layers but during a total eclipse the spectacle of the corona is a wonderful experience. Common sense dictates that the outside layer, the corona, should be cooler than the layer below it.

But it is not so, the corona is much hotter than the layer below and the explanation of how this can happen is still a matter of debate.

The various physical processes that provide the Sun with strong, frequently changing magnetic fields were pronounced not so long ago as 'both scientifically and literally shrouded in mystery.' To help unravel these processes clever instruments can now produce pictures of slices of the Sun's interior, rather like the scans that hospitals produce of the human body. We have seen that on the surface of the Sun different areas rotate with different speed. The same was found to be true also of at least a portion of the interior. Contemporary theory is that the differentially rotating, electrically conducting gases generate magnetic fields. The twists and turns of these fields then create more electric currents, with the result that even stronger magnetic fields are generated. The process is similar to that of a dynamo in an electricity-generating plant and is appropriately called the solar dynamo.

The current hypothesis is that a sunspot is born when the magnetic field is getting twisted into a rope-like structure on that part of the Sun and does not allow enough heat to reach the surface. Spots are therefore cooler than the surrounding area, and because they are cooler they appear darker. Spots die when the twists in the magnetic field unravel and heat can again travel outward from the Sun's interior unimpeded. Thomas Cowling, Professor of Mathematics at Leeds University in the middle of the last century, showed that the behaviour of spots could not be explained by a magnetic field growing and decaying because that would take many years, while spots undergo daily changes. Rather, the changes are the result of the existing magnetic fields moving around.

Until recently only those spots that were visible on the side of the Sun turned towards us could be studied. New methods, using ultrasound waves or ultraviolet radiation, can now provide information about what is happening to spots even when they are on the far side of the Sun, and turned away from us.

The discovery of sunspots opened the field of solar studies.

The Sun became a revolving globe. It had to be admitted that it is not uniformly bright and in due course the discovery of several other visible solar features followed. Here are a few examples: when there are plenty of spots there are usually also plenty of very bright *faculae* nearby (first noted by Scheiner). *Faculae* are brighter and hotter than the surrounding area, heat trapped below the sunspots escapes through them. *Plages* (meaning beaches in French) are other brighter areas. Prominences are clouds of gas, when seen against the surface they are called filaments. Occasional flares and coronal mass ejections (CMEs) are explosive outbursts. The visible surface of the Sun is described as granular and it is in constant movement, constant change. The changes are rapid when the Sun is very active and this is taken to be a sign of powerful physical processes going on in its bowels and on its surface.

We now know that the Sun is a constantly changing star, very much like some other stars. Some changes are of short duration: oscillations lasting five minutes and probably due to trapped sound waves have been recorded. Some changes are sedate and take place over long periods of time, some have a cyclical nature and some are violent. When the Sun changes the spots on its skin we can be sure that changes go on inside its body as well and the appearance and disappearance of the spots is an excellent indicator of these changes.

Schwabe's sunspot cycles are now called 'solar cycles.' The justification for this is that the sunspot cycle signals simultaneous changes in the Sun itself. How do we know? We know because many other observable features of the Sun also vary on average in a roughly eleven-year cycle and reach a maximum at the time of sunspot maximum. *Faculae,* flares and prominences, all show an eleven-year cycle; the corona reaches its maximum extension at sunspot maximum and even the size of the Sun varies according to this cycle. For easy reference, solar cycles are numbered. In the year 2000 cycle number 23 reached its maximum.*

When the solar cycle is at its maximum, and for some time before and after, the Sun is very active. Its magnetic properties

* Individual sunspots and sunspot groups also have their own numbers and classification according to various criteria.

undergo great changes. Because we see sunspots where magnetic fields emerge from the interior, the spots are for us one of the visible signs of the Sun's activity. At sunspot cycle maximum the Sun is a little less secretive than usual. It is the time when solar researchers are especially keen to monitor the conditions on the Sun, near the Earth and in the interplanetary space between the two. New data helps our understanding of what makes the Sun tick, or in other words, what the physical mechanism driving it is. Predictions and models can then be made and tested later to see if the theory on which they were based was right or wrong. Predictions of sunspot numbers can also be made on the basis of their past performance. Since our knowledge of either of these important factors is less than watertight, the forecasts are also less than certain. A slight glitch for instance, between October 1998 and April 1999 when sunspot numbers were very low, made scientists fear that their forecast about cycle 23 may have been wrong. The Sun did become satisfactorily more active afterwards and we can now be reasonably sure that the year 2000 was a year of sunspot maximum, as predicted.

That the activities of the Sun influence the Earth and life on Earth had of course, been recognized and experienced from the earliest of times: we can see the light with our eyes and feel the warmth of sunshine on our skin. Light and radiant heat certainly carry a large proportion of the energy we receive from the Sun and they are vital for maintaining life on our planet. But even a painful sunburn is not entirely the result of a cause that can be registered by our five senses. The red and peeling skin we suffer from when we neglect to smear on a sunscreen with high enough UV blocking factor is caused mostly by the ultraviolet rays. These rays we cannot see or feel on our skin, and we only rue their effect when it is too late. Radio waves also arrive from the Sun and can occasionally cause serious interference and impede telecommunications. Visible, infrared and ultraviolet light, radio waves and X-rays (also supplied by the Sun), all belong to the family of electromagnetic waves. Some we can experience directly with our senses, some others only indirectly, through their effects.

In addition to electromagnetic waves the Sun continually sends out streams of electrically charged (ionized) particles, and these also influence our environment. The solar wind which originates in the corona is a more or less steady flow, consisting mostly of protons, electrons and magnetic fields. Most solar wind particles are diverted by the magnetosphere and never reach us; some penetrate it and become trapped. The magnetosphere is the region in space where the Earth's magnetic field is in control. Shaped by the solar wind, it surrounds the Earth and acts as a protective shield. On the side facing the Sun the solar wind sets a boundary to the magnetosphere, and on the other side it sweeps it into space.

Solar wind particles that are constantly emitted by the Sun are not to be confused with cosmic rays. Cosmic rays have much more energy than the solar wind particles; they travel nearly as fast as light while solar wind speed is only somewhere between 300–800 kilometres per second. Solar wind is in some ways similar to our terrestrial wind; it is more densely packed together than the cosmic rays and behaves more like a gas.

The existence of the solar wind was suspected even before it was found and identified. The orientation of the tails of comets was for a long time vaguely attributed to pressure from the Sun. Then in 1942 when the tails of a newly appeared comet did not point exactly away from the Sun, the best explanation offered was that a flow of gas in space was blowing them off course. In 1947 we find in a book on sunspots: 'It takes little stretch of the imagination to believe that we shall ultimately find many ... [particles] in interplanetary space.'[1] And in 1958 the American astrophysicist Eugene Parker presented a theory of the solar wind. His prediction came true in 1962 when one of the Mariner 2 space probes aimed towards Venus, unequivocally detected the stream of particles originating from the Sun.

In addition to the solar wind, the Sun occasionally emits large bursts of matter and radiation from flares (noted by Carrington in 1859) and from coronal mass ejections (recognized only in the 1970s). These emissions are the results of violent upheavals and the particles in them travel faster than those emitted by a more quiet Sun. Sometimes they have enough energy to penetrate the magnetic shield. Our instruments are able to perceive these solar missives although our conscious senses are unable to do so. However, some of the invisible particles that manage to arrive here betray themselves even to our unaided eyes. When they interact with the atmosphere they produce the most beautiful auroras and thus make themselves known in an indirect way.

We are now certain that the magnetic storm of 1859 was caused by such unusually energetic, electrically charged particles and by electromagnetic radiation suddenly spewed out by the Sun from the area of the flare seen by Carrington. When this ejaculation reached our atmosphere it brought with it a taste of the Sun's magnetic field which interacted with that of the Earth. It has been calculated that the energy released into space during a convulsion of the Sun can be the equivalent of a hundred million hydrogen bombs.

Sunspots have no effect on the Earth's magnetic field; radiation and particles emitted by the Sun do. But there is an intimate connection between them. When there are few sunspots, the Sun is not very active. There are few flares and coronal mass ejections. The solar wind particles are generally not very energetic and can easily be diverted by the magnetosphere. But when there are many sunspots and the Sun is active, then it is occasionally liable to send out very energetic particles and radiation and these can cause disturbances with possibly devastating consequences for modern communications. We rely on a constant electricity supply, on telephone, radio and television, while satellites in the sky act as transmitters of radio waves and help aeroplanes reach their destination. A magnetic disturbance can affect all of these. If we could predict with any certainty the arrival of a magnetic storm, protective measures could be put in place at the right time. Unlike terrestrial weather predictions which have become quite reliable — and that is why better preparations for hurricanes can be made now — forecasts of magnetic storms are still largely unreliable. Although several spacecraft constantly take measurements in the interplanetary medium, there are occasions when we can still not be certain whether the particles from a detected emission are heading our way or not. Even if we know that we were in the path of the particles when they left the Sun, they could slow down on their way, accelerate, get deflected, and sometimes they can get entirely lost. The pattern of the magnetism they carry is also a factor of what kind of influence they are going to have.[2]

In 1859 particles ejected from the flare that Carrington noticed had enough energy to breach the magnetic shield. Agnes Clerke reported that during the magnetic storm the 'magnetic needle ... darted to and fro as if stricken with inexplicable panic.'[3] But it took time to figure out if these were just coincidences and if not, to come up with a valid explanation. For instance, in the early days of short wave radio communication, receivers blamed their own apparatus for complete blackouts, which were really due to a magnetic disturbance. They took their apparatus to pieces and tried to repair it. Only when others also reported the same problem it dawned on them that something more general was at work. By August 1972, when strong electric currents, induced by a sudden magnetic storm, overheated a huge transformer in British Columbia until it exploded, the explanation was already at hand. Similarly, in March 1989 discharges from flares and coronal mass ejections left much of the province of Quebec without power for nine hours. The north-eastern states of the US were also affected right down to New Jersey.

Solar storms often damage spacecraft and artificial satellites as well. They can damage their solar panels, interfere with their instruments and drag them off their orbits. In May 1998 the newspaper *US Today* carried the headline: 'Satellite's death puts millions out of touch' when the communication satellite Galaxy 4 failed catastrophically. In America 45 million pagers fell silent, meaning that doctors and fire crews could not be reached for emergency service. Ironically, solar particles sometimes debilitate the very spacecraft sent up to study the Sun! They also present a serious danger to the health of astronauts on long missions and could endanger their lives when they are outside a spacecraft.

During the Sun's active phase, bursts of radio waves reaching the top layer of the atmosphere can also cause disruption. Solar radio waves jammed British radar operations in an incident during World War II in February 1942, when three German battleships were not detected and were allowed to sail unmolested through the Channel. Although the Germans claimed that their own jamming operations prevented the ships from being detected, the real culprit was the Sun.

In 1932 the Sun was reasonably quiet. Karl Jansky, a radio engineer with the Bell Telephone Laboratories was studying static interference when he noticed radio waves from a source he could not locate at first. The waves seemed to come from outside the solar system and eventually he was able to point out the direction they were coming from. Jansky's serendipitous discovery heralded the beginning of radio astronomy. Had it been the time of a solar maximum, radio waves from the Sun would have jammed the waves coming from space and radio astronomy would not have been thought of for at least several more years. Solar radio waves are now monitored by ground-based and by satellite-borne radio telescopes.

How about auroras? That there was a connection between auroras and magnetism had already been known in the eighteenth century, well before the sunspot/magnetism connection was uncovered. In 1716 Halley theorized about auroras being caused by terrestrial magnetism but in southern England auroras are such a rarity that neither regular observation, nor experimentation was possible. In northern latitudes auroras are more frequent so it is no wonder that it fell to Scandinavian scientists to study them. A hundred years before the connection between sunspot cycles and cycles of magnetic variation was announced, the Swedish astronomer Andreas Celsius (who devised the centigrade temperature scale) noticed the erratic movements of the magnetic needle when an aurora was seen in the sky. To make sure of the connection, Celsius asked his brother-in-law Olof Hiorter to take readings of the variation, which he duly did. Hiorter took over 6000 readings, almost every hour for a whole year between 1741 and 1742! The result was clear: whenever an aurora danced in the sky, the magnetic needle danced in its case.

THE ENIGMA OF SUNSPOTS

John Canton, an English schoolteacher, performed a similar marathon. All in all, between 1756 and 1759 he took 4000 compass readings spread out at various times of the day. (After self-registering instruments were invented sacrifices like those made by Hiorter and Canton became a thing of the past in the physical sciences.) Having previously shown experimentally that a magnet loses some of its strength when heated, Canton accounted for the diurnal magnetic variation by the changes in the amount of sunshine falling on the Earth throughout the day. By analogy Canton surmised that change in subterranean heat was the cause both of irregular changes in terrestrial magnetism and of auroras.

Meanwhile the suspicion that the Sun, or the Moon, might be responsible for the appearance of auroras was gradually gaining ground. At the beginning of the twentieth century Svante Arrhenius, the Swedish electrochemist, was interested in auroras and their possible periodicity. He was convinced that ultraviolet rays from the Sun and maybe even heat and light are partial causes for the polar lights. He held a direct solar electromagnetic effect to be not only unproven but also extremely unlikely.

Why are auroras more frequent near the north and south poles? Normally the solar wind particles are not especially energetic and cannot breach the magnetic shield, except in an area close to the polar regions where the shield is relatively weak. There they manage to sneak through. These solar particles excite the molecules of the atmosphere which then produce the light fantastic. That is why auroras can usually only be seen in the far north and the far south.

But the more energetic particles sent out by the active Sun can force their way through at lower latitudes as well. After the 1989 magnetic disturbance the most beautiful auroras danced merrily in the sky as far south as Texas in the United States and as far north in the southern hemisphere as Queensland in Australia. In May 2000, close to the solar cycle maximum, beautiful auroras could be seen in various parts of the world without having to make a trip to the Far North. Although it is still not possible to predict exactly when and where auroras will brighten the sky after a solar emission, there is always the possibility that an aurora will appear and those who live in the northern hemisphere and are keen to experience an auroral display should then keep their eyes in a northernly direction on the night sky.[4]

∿ 16 ⋍

Tenuous Links

The scientific and lay community had been used to the unpredictable nature of sunspots for such a long time that it was at first reluctant to accept their cyclic nature. But when Wolf finally won his case there was a sudden rush to search for similar cycles here on Earth. The supposition was that cyclical changes on the Sun would manifest themselves in cyclical changes on Earth and once these were found, forward planning would become so much easier. Forecasting is important for life on Earth. The farmer is eager to know what the weather will bring, the fisherman the likely catch, and citizens living in Florida what they can expect in the hurricane season. Past regularities allow us to make such predictions. Of course, predictions are just that: events that are likely but not certain to occur.

When a search is on, people are sure to find what they are searching for, and this is exactly what happened. Many eleven-year cycles were found and immediately connected to the sunspot cycle. We have seen earlier that in previous times the amount of rainfall had been linked to sunspots. Now the hunt was on for rainfall cycles. Herschel had related temperature to sunspots. Now temperature cycles were hot news, and we have seen that Douglass found them in tree rings.

Nor did many other natural phenomena escape attention. Fish catches in Russia, insect populations in Britain, mammal populations in Canada, the amount of annual snowfall in Massachusetts, bird migrations: these were all examined and correlated to sunspot numbers. Tropical cyclones in India, hurricanes in America and thunderstorms in temperate regions were also supposed to vary according to the solar cycle. Cycles of drought were uncovered. Cycles of the water level in the American Great

Lakes and in Lake Victoria in Africa, and cyclical variations in the flow of the Nile were announced. The quality of wine vintages was investigated and it was hoped that their cyclical changes could stand in for the amount of sunshine the vines received and the summer temperatures during the ripening of the grapes.[1]

Human activities did not escape the attention of cycle searchers either. Business cycles, economic conditions and stock exchange crashes were all correlated to the sunspot cycle. This is not as crazy as it seems. All life is influenced by the daily cycle of light and darkness and we now accept the existence of Seasonal Affective Disorder (SAD), caused by lack of sunshine in the British winter. During the sunspot cycle the amount of the Sun's rays reaching us varies slightly and so does their composition. So there is no reason to discard out of hand the idea that the solar cycle has some influence on human physiology or on the human psyche, which in turn influences the economy. In some countries (such as for instance in Hungary), people will blame 'sunspot outbursts' when they feel unusually nervous or excited.

One cycle governed by the solar cycle exists without a doubt: a roughly eleven-year cycle in long distance radio signal reception. When radio broadcasting began at the turn of the nineteenth century, there was no explanation of why high frequency, short radio waves which travel in a straight line and are not able to follow the curvature of the Earth, still manage to reach faraway destinations. Oliver Heaviside and Arthur Kennelly thought that they might be reflected from an ionized layer in the upper atmosphere. Twenty years later Sir Edward Appleton, who got the radio bug as signals officer with the British Royal Engineers in the First World War, provided experimental evidence for not just one but several such layers. These layers, which make up the ionosphere, are created by ultraviolet rays and by X-rays arriving from the Sun, stripping electrons from atoms and molecules. The strength of ionization depends of course on the sunshine and consequently radio reception varies with the time of day and with the seasons. It varies in step with the solar cycle as well. During low

THE ENIGMA OF SUNSPOTS

solar activity, when there are few sunspots, the ionosphere is weak (i.e. there are few ionized particles) and cannot support the transmission of very high frequency radio waves from sender to receiver. Broadcasters have to alter their wavelengths and sometimes relay transmitter stations have to be called in to help. On the other hand, at solar maximum the ionosphere will support shorter waves but disturbances due to flares and other eruptions become more frequent. Electromagnetic waves from a flare can cause fade-out seconds after the eruption and some time later when solar wind particles arrive, total blackout can be the result. The connection between magnetic storms and radio reception (or rather non-reception) was also made by Appleton when in Norway he experienced annoying blackouts while at the same time the Northern Lights were exceptionally beautiful. Appleton received the Nobel Prize in 1947 for his investigations of the ionosphere.

European nations still send out about 7000 hours of public and commercial broadcasts on short wave frequencies every day, so the correlation between solar activity and the behaviour of the ionosphere is of vital interest. The World Radio TV Handbook (WRTH in short, the bible of short wave broadcasting) has a special section on sunspots and on the forecasting of space weather.

The search for cycles is still going on. There was great excitement when in the 1980s eleven-year cycles were found in 680-million-year-old sedimentary rocks near Adelaide in South Australia. Financial backing was arranged for George Williams to make a thorough study of them. Analysing the sediments going back for over 1000 years, he thought that he could relate their cycles to the sunspot cycles. If correct, this would be a unique and truly useful historical record. But doubts arose and it seems now that the rocks show the ebb and flow of the tides rather than the abundance or dearth of sunspots.

A less serious correlation found in 1989 supposes that the prediction of winning numbers in the lottery varies according to the number of sunspots. Recently an astrologer claimed to have found a correlation between the solar cycle and hormone production in the human body. (As explained above, this sounds quite plausible and would be worth scientific investigation.) Reverting to Kircher's idea that connection with the Sun instead of the stars should be sought, he developed a theory that astrological differences between people are caused by solar activity at conception. (Not at birth!) Doing some esoteric number crunching the same astrologer discovered a new, 187-year-long solar cycle matching certain numbers that had special importance for the Mayans.

Another project, conducted by 'Bible Code' researchers, statistically analyses letters and words in the Old Testament. The researchers claim that predictions of historical events can be discerned and that many of the deciphered codes refer to solar cycles and catastrophes awaiting us at the time of a solar maximum.

Conventional science does not escape suspicion either when too many different cycles are found and correlated with too many different events. Besides the approximately 11 year sunspot cycle the Sun is supposed to exhibit a 27 and a 154 day cycle, the 22 year cycle of magnetic reversal found by Hale, also a Gleissberg cycle of about 80 years and a 200 year cycle called the Seuss cycle. The 27 day cycle corresponds to the Sun's rotation and although, as Arrhenius pointed out, it is associated with auroras, its association with the weather is more contentious. Is there a connection between terrestrial events and the solar cycle? One thing is certain: the connection between changes in the Earth's magnetic field and the sunspot cycle, or to put it more accurately, the solar cycle, is real. As to the rest, our knowledge is rather hazy. One of the problems is that many of the cyclical events mentioned above are local, or if not strictly local, they apply only to a small area. Also, while many of the studies seem to be accurate enough for a limited time interval, they become hopelessly wrong when extended. Many of the findings turned out to be contradictory and some were the results of faulty data or faulty statistics. A recent publication by experts in the field is cautiously pessimistic: 'In many instances, the conclusions are discouraging, but a sufficient number of positive findings should encourage further work.'[2]

Could changes in solar radiation have long term effects on the climate? In the last twenty years warnings of global warming have become increasingly more frequent and more urgent in tone. The important question — whether climate change has anything to do with the Sun and its sunspots, or is entirely the result of human activity — is still debated. If global warming is completely man made — caused by our greenhouse gases — then we can, and ought, to do something about it. On the other hand, if most of it is due to the Sun warming us more than in previous ages, or to some other reason outside our influence, then there is not much we can do to avert it. In that case our job is limited to make the best of it by careful planning.

Climate is not easy to define. Temperature, rainfall, snowfall, wind, storm activity and cloud cover are all part and parcel of climate. All can be influenced by the Sun, by terrestrial events and by human activity. The term global warming often includes all the baggage that belongs to climate although it should refer only to one of these factors: change in the mean temperature.

There is no doubt that geological and human factors have an influence on the temperature. On a very long-term scale, measured in tens of thousands of years, changes in the tilt of the Earth's axis may have caused the Ice Ages. (This is the famous Milankovitch Hypothesis.) For changes of the natural kind on a shorter time scale, volcanic eruptions spring to mind. Ash and dust particles blown into the air block out the Sun and have a cooling effect. The 'year without a summer' of 1816 was caused by an eruption of Mount Tambora on the island of Sumbawa, now in Indonesia. Smoke made by humans is similar and coal fires were certainly effective in cooling London in the nineteenth century. These particles are now described as 'aerosols.' Changes in the flora might also have an effect. Greenhouse gases, both of 'natural' and of 'human' origin (inverted commas indicate that there is nothing unnatural about human activity) are frequently blamed, with carbon dioxide and methane cited as the major culprits. Both are produced by biological processes and also by the burning of fossil fuel.

There can be no doubt that the amount of radiation reaching

the Earth influences the climate, in fact without it there would not be any climate at all. In much of our story astronomers were handicapped by having to observe the Sun and the radiation we receive from the ground, at best from high mountain tops, from balloons or aircraft, but always through the atmosphere. Now spacecraft that carry better instruments than ever before can give us more precise answers. These instruments perform a variety of measurements and one of those is the monitoring of what used to be called the 'solar constant.' This was defined as the total energy in all wavelengths the Sun radiates towards the Earth, the measure of it being the amount crossing a unit area in unit time at the Earth's mean orbital distance. Unsurprisingly it was named the solar constant because for a long time it was believed to be constant, never changing. Now it is better referred to as 'irradiance.'

The first experiments to measure the radiation reaching us from the Sun date back to the 1830s. Think of heating the house in winter. If the temperature in the house remains constant then the amount of heat generated by the boiler equals the amount of heat lost through the roof, the walls and the windows. The same principle is used for measuring the heat we receive from the Sun. The first instruments were insulated blackened copper vessels with two small holes. They had a thermometer fixed inside so that the Sun shone onto the bulb through one of the holes and readings could be taken of the thermometer sticking out the other end. When the thermometer reading stayed constant then the heat gained from the Sun equalled the heat lost from the whole instrument and this could be calculated. Of course, this measurement took place beneath the cover of the atmosphere which absorbs some of the radiation. Later the solar scientists became mountaineers and carried their instruments up high peaks to eliminate as much of the influence of the atmosphere as possible.

Modern instruments carried by spacecraft still consist basically of a box painted black inside, but their whole apparatus is now a great deal more complex and sophisticated than a thermometer in a bowl. They are unhampered by weather and the

THE ENIGMA OF SUNSPOTS

intervening atmosphere and are always pointed automatically in the right direction. The cavity, heated by electricity, is maintained at a constant temperature and the current needed for this is measured.

In the 1970s efforts were made to measure the radiation received by the Earth and compare it to the radiation sent out into space. This was the Earth Radiation Budget experiment of the *Nimbus* satellites. The *Solar Maximum Mission (SMM)* was launched in 1980, but a delay caused it to just miss the maximum of solar cycle 21. It carried the first *Active Cavity Radiometer Irradiance Monitor*, known as *ACRIM*. The *SMM* is famous because when it became debilitated a manned shuttle was sent up, made a rendezvous with it and the astronauts repaired the damaged satellite in space. Finally a magnetic storm damaged it beyond repair and in 1989 its life ended with a plunge of the debris into the Indian Ocean. In not quite ten years the *Solar Maximum Mission* recorded 12,500 flares. This shows that flares are not such a rarity as they were once thought to be. The *UARS* (*Upper Atmospheric Research Satellite*) carrying *ACRIM II*, was launched in 1991. That same year saw the launch of the Japanese *Yohkoh* (sunbeam) mission. *SOHO* (*Solar and Heliospheric Observatory*) was launched in 1995, *ACE* (*Advanced Composition Explorer*) in 1997 and *ACRIMSAT* in December 1999.

Surprise, surprise, the solar constant as measured by spacecraft borne instruments turned out to be not constant at all, but changing with the sunspot cycle. True, within a cycle the whole variation is about one tenth of one percent, which is not much, but definitely more than zero. There are periods when more radiation reaches the Earth even though more of the face of the Sun is covered with spots because brighter features compensate for the relative darkness of the spots. Historical research points to periods when it could have been the other way round and this might be the explanation why Herschel and Gautier arrived at diametrically opposite conclusions. Radiation changes are not uniform, in general it can be said that the shorter the wavelength the more it varies with the solar cycle. We have seen the importance of this in the creation of the ionosphere.

Variation within a solar cycle is not great and in any case if the cycles are more or less equal then they cannot exert any long term warming or cooling influence. If the Sun shines brighter at sunspot maximum then it will be dimmer at the minimum. When a whole cycle has gone by, the warming and cooling more or less cancel each other out.

But solar cycles are not uniform: they vary both in their intensity and their length. Some cycles are as short as eight years and some are thirteen years long or even longer, the eleven year cycle is only an average. Also, during some cycles the Sun is very active with many relatively darker and brighter areas, during others the Sun is more quiet.[3] During periods of extended sunspot minima the change in the Sun's output could also have a definite effect on the climate.

To find a long-term connection, if any, between sunspot activity and temperature, we need to examine past records of both. It is difficult to assemble a picture of the present average temperature but when it comes to reconstructing temperatures of remote ages, needed for comparison, the task becomes even harder. Nevertheless, people have made an attempt to reconstruct past temperatures. Extant records are few and unreliable and a medley of different methods had to be used, ranging from measuring the thickness of tree rings, to studying the varieties of shells on the ocean floor. Apart from the last two centuries, the influence of human activity can be largely disregarded, and any variation must be due to the Sun and to other natural causes. Reconstruction of twentieth century temperatures relies on actual measurements taken. It shows a slight decrease in the middle of the century (those who are old enough will remember the cold winters in the 1960s and 1970s when we were warned of the coming of a new Ice Age) with a larger increase afterwards.

Does solar activity influence the temperature? We can simply take the number of sunspots as an indicator: when the Sun is more active then it has more spots on its face. Reliable sunspot data is available for the last 150 years. Nowadays we can also watch the behaviour of other stars, similar to the Sun, and draw comparisons. This allows us to go backwards and forwards in time because some stars are at an earlier stage of development than ours while others are at a later stage. Such a study resulted in the conjecture that the Sun was dimmer during the Maunder Minimum than it is now.

Another method of reconstructing the Sun's activity in time, as we have already seen, is the measurement of the amount of carbon-14 isotope in tree rings. Cosmic rays, whose intensity varies with the Sun's activity, produce another isotope as well, beryllium-10. This is best found in ice cores from Greenland and is considered a better indicator because it is less dependent on living organisms. But beryllium-10 must also be used with caution because it does not always agree with past observations, for example, it shows the Dalton Minimum, but not the Maunder Minimum.

All these measurements point to an increasingly active Sun during the last few cycles. Another indicator of solar activity has also increased: this is the variation of the geomagnetic field. Remember, there was a connection between geomagnetic field variation and sunspots, a discovery claimed by Wolf, Sabine and Gautier. More sunspots indicate a more active Sun. A more active Sun sends out more charged particles that interact with the geomagnetic field. Scientists who scrutinized the records going back to 1868 found that on the whole geomagnetic activity has increased since then. Reasoning backwards, one can say, that if geomagnetic activity has increased, then the Sun's activity must have also increased in the last 150 years. Actually, the rate at which magnetism leaves the Sun has risen by a factor of 1.4 since 1964 and by a factor of 2.3 since the beginning of the twentieth century. The index of geomagnetic activity seems to be a good proxy indicator of not only solar variability but also of recent climate change. In our time the Sun has become increasingly more active and an active Sun radiates more: as a result the temperature rises.

Trying to link sunspot activity and terrestrial temperature produced some unexpected and sometimes questionable correlations. The Danish scientists E. Friis-Christensen and

K. Lassen noticed a correspondence between the length of the solar cycles and temperature in the northern hemisphere. They pointed out that for the last hundred years or so, the shorter the cycles were, the warmer the temperature and this could possibly be extended backwards in time. What sort of physical mechanism could account for this? So far none has been found. The correspondence may be a coincidence, especially since recently it was found that it has not held strong for the last twenty years.

Another lead D. V. Hoyt followed up has to do with the ratio of the area of the sunspot nucleus to the area of the surrounding penumbra. Apparently this also tracks the temperature deviation. So does the rate at which sunspots decay. These relations are getting attention not because of our interest in sunspots *per se,* but because we hope to be able to use them as proxy measurements for the Sun's output and they might give us an insight into how the Sun influences the climate on a longer time scale.

Recently yet another possible explanation was put forward for global warming. A more active Sun emits more energetic particles. We have discussed some of the effects of these particles: they limit how far the Earth's magnetic field extends, they cause the auroras, and they drive away the cosmic rays. It was this last effect that drew the attention of the Danish scientists H. Svensmark and E. Friis-Christensen. The basis of their theory is that cosmic rays have a hand, so to speak, in the formation of clouds. If more cosmic rays reach the atmosphere then more clouds are produced while fewer cosmic rays result in less cloud cover. Cloud cover acts in two diametrically opposite ways: it prevents some of the heat escaping from the Earth so has a warming effect, but it also prevents the rays of the Sun from reaching us, resulting in a cooling effect. On balance it has been found that the cooling effect predominates.

So to recap. A more active Sun means more solar wind; which means fewer cosmic rays, less cloud, warmer Earth? At some stage it has even been suggested that global warming might be entirely due to this effect. This theory was a few years ago vigorously popularized but also criticized. It is now

THE ENIGMA OF SUNSPOTS

again under consideration and is about to be experimentally tested.

According to this scenario, global climate during the Maunder Minimum must have been much cloudier (and cooler) than it is today. Looking at Dutch landscape paintings might convince us that, at least in Holland, the weather was indeed cloudy much of the time. However, this is what Eddy had to say about cloud cover during the Maunder Minimum:

> There is no evidence of unbroken overcast. Astronomers are neither so mute nor so long-suffering that they would have kept quiet through year after year of continuous, frustrating cloud cover.[4]

The so-called Little Ice Age, years of extreme cold when the River Thames froze over and frost fairs were held on its ice, and which partly coincided with the Maunder Minimum, is also often quoted as proof of a colder climate in Europe at the time. There are dissenting voices though. A dimmer Sun could have accounted for the cold. Another suggestion is that climate in the Middle Ages between 900 and 1250, was distinctly warm and started to cool in the thirteenth century, well before the Maunder Minimum, that is, when the Sun was still active. Our current warming trend might still be part of the recovery from the Little Ice Age. And in any case, did the cold weather extend to the southern hemisphere? 'Anecdotal and regional accounts give a confusing and contradictory picture of the Maunder Minimum climate' is the verdict in a recent publication by solar scientists.[5]

At the moment all we have are surmises and it is pretty unlikely that we shall ever stumble on four-or-five-hundred-year-old reliable meteorological records for the whole planet. Already in 1894 Maunder wished he could find one: 'How valuable would be a few of those ponderous volumes of rainfall and temperature that, when published in modern observatories and in these days, we are apt to look upon with scorn.'[6]

We have learnt our lesson and changed our attitudes. Figures of rainfall and of temperature are not looked upon with scorn any more; on the contrary, they are kept as valuable records. Still, the verdict at present has to remain that 'neither climate, nor solar variability are sufficiently well defined, either spatially or temporally, nor their causes adequately understood.'[7] Increasing solar and human activity both contribute to global warming but in what proportion is still unknown. Enlightenment may still come from dark spots on the busy old fool, our unruly Sun.

Appendix 1
Some Literary Allusions

It is a pity that we are much less poetically minded than the ancient Chinese were in describing sunspots. We saw in Chapter 2 how they saw everything from a flying magpie to a plum in the Sun's surface. At their first discovery Europeans simply called the spots *maculae* in Latin, meaning spots but also suggesting blemishes. A dark spot on the Sun was considered a blemish by those who believed the Sun to be the eye of the universe and by those who believed that heavenly bodies were perfect. A blemish it certainly was for Milton when in his *Paradise Lost* he made Satan fall into the Sun:

> There lands the Fiend, a spot like which perhaps
> Astronomer in the Sun's lucent Orb
> Through his glazed Optic Tube yet never saw.

In the seventeenth century the spots were still likened to the 'tail of a scorpion' or a 'melon-seed' but all these descriptions have by now given way to the simple and unimaginative 'sunspot.'

Being unexplained and exotic, sunspots lent themselves well to social and political satire. They appear as religious symbols in Milton's poem. Andrew Marvell, poet, politician and secretary to Milton, addressed Charles II in his political satire *The last instructions to a painter,* like this:

> So his bold tube, man to Sun applied
> And spots unknown to the bright Sun descried
> Showed they obscure him, while near they please
> And seem his courtiers, are but his disease
> Through optic trunk the planet seemed to hear
> And hurls them off e'er since in his career.

The last two lines, written in 1667, have been interpreted as a reference to the Maunder Minimum.

The Sun as a place for unwanted creatures and its spots as malevolent agents also feature in the 1816 satire *Napoleon and the spots in the sun; or; the R–t's waltz; and who waltzed with him — and where.* R–t in the title of the satire refers of course, to the Prince Regent, later George IV whose pleasure loving and lavish lifestyle and unconventional domestic arrangements (to say the least) made him highly unpopular.

The waltz, an exciting new dance, first arrived in England in 1812 and at once it was objected to and ridiculed in equal measure. Why? Because the partners' close hold, with the man putting his arm around the woman's waist, was previously unheard of. Nevertheless, at the Congress of Vienna, where world peace or rather, the division of power in Europe was discussed, the waltz was danced with gusto. On July 16, 1816, *The Times* 'remarked with pain' that the 'indecent foreign dance called the Waltz was introduced ... at the English court ... it is quite sufficient to cast one's eyes on the voluptuous intertwining of the limbs and close compressure of the bodies, in this dance, to see that it is far indeed removed from the modest reserve which has hitherto been considered distinctive of English females ... we feel it a duty to warn every parent against exposing his daughter to so fatal a contagion.' The leader writer felt that the Royal Family and the aristocracy were giving a bad example to the middle classes.

Now that we have put the matter in context, the *real* cause of sunspots as presented by the author, a certain Syntax Sidrophel, can be revealed. It is none other than Napoleon who resides in the Sun because he had nowhere else to go after his downfall.

> Peep through this smok'd glass and you'll see
> The spots are his body and limbs
> As plain as plain as ever can be:
> The bright parts are what he don't hide
> By his head, arms, legs, body, and shoulders;
> If he should put on his great-coat,
> It will seem an Eclipse to beholders
> And puzzle great Herschel himself

Napoleon, as a collection of sunspots, blocks the sunshine and causes the unusually wet and cold weather. There is even a hint as to the origin of the water dribbling from the sky. Someone must be sent up to stop Napoleon making mischief. The best person for this would be the Prince Regent. He is duly dispatched in the hope that the two will waltz together until they both get dizzy and 'the chance is they both will fall out! And the spots in the Sun disappear.'

The year 1816 was indeed unusually cold and wet, needless to say not due to Napoleon blocking the Sun. News that Russia had an exceptionally hot summer put paid to the rumour that the Sun's energy supply was getting exhausted. We now think that the miserable weather was due to the 1815 eruption of the volcano Mount Tambora. The year 1816 was also a year of sunspot maximum, just like 2000.

A one-act French vaudeville performed in 1816 poked fun both at Lalande and at human behaviour. The well-named M. Desastres, barber surgeon and amateur astronomer, regularly declares that the end of the world is imminent. His prophecies are usually disregarded. But on this occasion during a conversation with Lalande while giving him a shave, the famous astronomer agreed with him. Desastres must also be believed because a peep through his telescope shows a spot looking like a great beast in the Sun. Getting ready for the inevitable, long neglected matters are put right, marriages are hastily arranged and debts re-paid. The end of the world is duly postponed when a cat jumps out of the telescope and it turns out that its tail was mistaken for the sunspot.

Appendix 2
Safe Observation of the Sun

The following are only a few hints on how to observe sunspots without ill effects. Before embarking on a voyage of discovery the reader is advised to consult *Guide to observing the Sun* published by the Solar Section of the British Astronomical Association, or *Observing the Sun* by Peter Taylor (Cambridge, Cambridge University Press, 1991). Useful information can also be found in: Medkeff, J. 'A beginner's guide to solar observing.' *Sky and Telescope,* vol.96, June 1999, pp.122–126.

The basic safety rules, which have to be strictly observed are:
Never ever look into the Sun. Not even for a split second, however tempting it might be to screw up your eyes and take a peep.
Never leave a telescope unattended.
Cap or remove the finder-telescope.

There are two methods for observing the Sun:

1. Directly through a telescope.

It should be noted that this method is still regarded as dangerous by some authorities.
Fit a safe filter, one that is specifically sold for the purpose, to the aperture. Fit it securely so that wind or a sudden jolt cannot dislodge it. Inspect the filter for any imperfections every time before use and discard if not perfect. Hold up a card and point the telescope towards the Sun by looking at its shadow. When the shadow is minimized, the telescope is aimed at the Sun.

2. By projection.

This is the safest way and although a little more laborious than direct observation, its advantage is that a permanent record of the spots can be drawn on a piece of paper. Be careful not to let the instrument get too hot. Do not use an eyepiece that is cemented together. Fix a card on an adjustable stand or ask a helper to hold it. Aim the telescope towards the Sun by the above described shadow method. Focus the telescope by looking at the image of the Sun on the card. Draw a circle on a piece of paper and attach this paper to the card. Then, for best results, follow Galileo's method:

> By moving the paper towards or away from the tube, I find the exact place where the image of the Sun is enlarged to the measure of the circle I have drawn. This also serves me as a norm and rule for getting the plane of the paper right, so that it will not be tilted to the luminous cone of sunlight that emerges from the telescope. For if the paper is oblique, the section will be oval and not circular, and therefore will not perfectly fit the circumference drawn on the paper. By tilting the paper the proper position is easily found, and then with a pen one may mark out the spots in their right sizes, shapes, and positions ... But one must work dextrously, following the movement of the sun and frequently moving the telescope, which must be kept directly on the sun. Also in order for the spots to be seen distinctly and with sharp boundaries, it is good to darken the room ... so that no light enters except through the tube, or at least to darken it as much as one can by fitting a rather large paper upon the tube to shade the other paper upon which one intends to draw, thus preventing any other light falling upon that paper. (Galileo 1957, pp. 115–16.)

Less dexterity is required if the telescope automatically follows the movement of the Sun.

Personalities

Abu Bakr Muhammad ibn Yahya (Avempace, also known as Ibn Bajjah) (*c.*1095–1138/39) Spanish Arab philosopher.

Appleton, Sir Edward Victor (1892–1965) English physicist.

Arago, Dominique François Jean (1786–1853) Director of the Paris Observatory.

Aristotle (384–322 BC) Ancient Greek philosopher.

Avempace, *see* Abu Bakr Muhammad ibn Yahya.

Averröes, *see* Ibn Rushd.

Bernoulli, Jean, *the younger* (1744–1807) Swiss astronomer.

Bode, Johann Elert (1747–1826) German astronomer.

Brahe, Tycho (1546–1601) Danish astronomer.

Bunsen, Robert Wilhelm (1811–99) German chemist.

Canton, John (1718–72) Mathematics teacher.

Carrington, Richard Christopher (1826–75) English amateur astronomer.

Cassini, Giovanni Domenico (1625–1712) Italian astronomer, later living in France.

Castelli, Benedetto (1577–1644) Galileo's student. Benedictine monk and mathematician.

Celsius, Anders (1701–44) Swedish astronomer.

Clerke, Agnes Mary (1842–1907) Historian of astronomy.

Copernicus, Nicolas (1473–1543) Polish astronomer.

Cowling, Thomas (1906–90) Professor of Mathematics at Leeds University.

Cysat, Johann Baptist (1586–1657) Jesuit, born in Lucerne, professor of mathematics in Germany.

De La Rue, Warren (1815–89) British astronomer.

Derham, William (1657–1735) English amateur astronomer.

Descartes, René (1596–1650) French philosopher.

Douglass, A.E. (1867–1962) Astronomer at Lowell observatory, Arizona. Discoverer of tree ring dating.

Fabricius, Johann (1587–1615?) Author of first publication on sunspots.

Faye, Hervé (1814–1902) Professor of astronomy at the École Polytechnique.

Flamsteed, John (1646–1719) First Astronomer Royal.

Flaugergues, Honoré (1755–1835) French amateur astronomer.

Fraunhofer, Joseph von (1787–1826) German optician.

Galileo, Galilei (1564–1642) The first to study the sky with a telescope.

Gassendi, Pierre (1592–1655) Priest, French astronomer.

Gautier, Alfred (1793–1881) Director of the observatory in Geneva.

Gualterotti, Raffaello (1543–1638) Florentine painter and poet.

Guericke, Otto von (1602–86) German scientist, mayor of Magdeburg.

Hale, George Ellery (1868–1938) American astronomer.

Halley, Edmund (1656–1742) Astronomer, famous for predicting the return of a comet.

Harriot, Thomas (1560–1621) English scientist.

Henry, Joseph (1797–1878) Professor of natural philosophy at Princeton. First secretary of the Smithsonian Institution.

Herschel, Caroline Lucretia (1750–1848) Astronomer of German extraction.

Herschel, Sir John Frederick William (1792–1871) English astronomer.

Herschel, Sir William (1738–1822) German musician, settled in England. Discovered the planet Uranus. Also discovered infrared radiation from the Sun.

Hevelius, Johannes (1611–87) Polish astronomer.

Hiorter, Olof Peter (1696–1750) Professor of astronomy at Uppsala in Sweden.

Hodgson, Richard (1804–72) Amateur astronomer.

Hooke, Robert (1635–1703) Natural scientist.

Horrocks, Jeremiah (1619–41) English amateur astronomer.

Humboldt, Friedrich Wilhelm Heinrich Alexander (1769–1859) German naturalist.

Huygens, Christiaan (1629–95) Dutch mathematician and scientist.

Ibn Bajjah (Avempace), *see* Abu Bakr Muhammad ibn Yahya.

Ibn Rushd (Averröes) (1126–98) Islamic philosopher, interpreter of Aristotle.

Jansky, Karl Guthe (1905–50) Discovered radio waves from space.

Janssen, Pierre Jules César (1824–1907) Director of the observatory at Meudon, France.

Kant, Immanuel (1724–1804) German philosopher.

Kepler, Johannes (1571–1630) German astronomer. Described the path of the planets.

Kircher, Athanasius (1601–80) German Jesuit of universal erudition.

Kirchhoff, Gustav Robert (1824–87) German chemist, university professor.

Lalande, Joseph Jerome (1732–1807) Director of the Paris Observatory.

Laplace, Pierre Simon, Marquis de (1749–1827) French mathematician and astronomer.

Lockyer, Sir Joseph Norman (1836–1920) Director of the Solar Observatory at Kensington. Discovered the element Helium.

Malapert, Carol (1581–1630) Jesuit mathematician.

Marius (Mayer), Simon (1573–1624) German mathematician and astronomer.

Maunder, Annie Scott Dill (1868–1947) British astronomer.

Maunder, Edward Walter (1851–1928) Astronomer at Greenwich.

Nasmyth, James (1808–90) Engineer, inventor of the steam hammer.

Newton, Sir Isaac (1641–1727) English mathematician and physicist. Classical physics and astronomy are based on his theory of gravitation and on his laws of motion.

Peiresc, Nicolas Claude Fabri de (1580–1637) French amateur astronomer.

Raleigh, Sir Walter (1552–1618) English courtier.

Rigaud, Stephen Peter (1774–1839) Professor of astronomy at Oxford.

Robertson, Abraham (1751–1826) Professor of astronomy at Oxford.

Scheiner, Christoph (1575–1650) Jesuit priest. Mathematician and astronomer.

Schwabe, Heinrich (1789–1875) German amateur astronomer.

Secchi, Pietro Angelo (1818–78) Italian astronomer.

Spoerer, Gustav (1822–95) German astronomer.

Tarde, Jean (1561?–1636) Vicar-general of Sarlat in France.

Theophrastus (*c.*372–*c.*287 BC) Ancient Greek natural scientist.

Virgil, Publius Vergilius Maro (70–19 BC) Roman poet, author of the *Aeneid.*

Wiedeburg, Johann Ernst Basilius (1733–89) Professor at Jena.

Wilson, Alexander (1714–86) Scottish astronomer.

Wolf, Johann Rudolf (1816–93) First director of the observatory in Zürich, Switzerland.

Wollaston, William Hyde (1766–1828) English chemist.

Zach, Franz Xaver, Baron von (1754–1832) Founder of the observatory at Gotha, Germany.

Zeeman, Pieter (1865–1943) Dutch physicist.

Endnotes

Chapter 1

1. Parker 2000, p.26
2. Newton 1961, vol.3, p.153.
3. Huggins, 1889, p.249. Sir William Huggins had an observatory at Tulse Hill in south London. He was the first to find the red shift in stellar spectra.

Chapter 2

1. Sayce n.d., p.53–54.
2. Wittman & Xu 1987.
3. Yau & Stephenson 1988, p.178.
4. Theophrastus 1894, p.56.
5. Virgil 1986, vol.1, p.111.
6. Einhard the Frank 1970, p.76.
7. Welsh 1904, vol.6, p.462.
8. Welsh 1904, vol.6, p.467.

Chapter 3

1. Aristotle 1984, vol.1, p.450.
2. Aristotle 1984, vol.1, p.477.

Chapter 4

1. Marius 1916, p.371.
2. Galileo 1989, p.37.

Chapter 5

1. Galileo 1957, p.116–117.
2. Fabricius 1611, C3–C4. My translation.

Chapter 6

1. Hooke 1676, p.3.

2. Galileo 1957, p.99.
3. Galileo 1953, p.50.
4. Galileo 1957, p.103.
5. Galileo 1957, p.106.
6. Galileo 1953, p.346.
7. Guiducci 1960, p.24.
8. Galileo 1953, p.345.
9. Drake 1978, p.337.
10. Braunmühl 1891, p.23.
11. Braunmühl 1891, p.58.
12. Arago 1855–58, vol.1 p.427.
13. Galileo 1953, p.48.
14. Galileo 1953, p.54.
15. Galileo 1953, p.54.
16. Galileo 1953, p.353.
17. Galileo 1953, p.355.

Chapter 7

1. Elizabeth I 1584.
2. Harriot 1588, p.C3 verso.
3. Bartha 2000, p.92.
4. Shirley 1983, p.14.
5. Shirley 1983, p.25.

Chapter 8

1. Arago 1842, p.463. My translation.
2. Gassendi 1657, book 4, pp.62f.
3. Hooke 1674, a2.

Chapter 9

1. Galileo 1953, p.51.
2. Marius 1916, p.373.
3. Burton 1621, p.329.
4. Descartes 1983, p.137.

5. Huygens 1689, p.219.
6. Wolf 1861, p.27–28. My translation.

Chapter 10

1. Kepler 1953, p.899. My translation.
2. Clerke 1902, p.125.
3. Maunder 1900, p.55. Flamsteed received only £100 per annum, and had to provide his own instruments. To supplement his stipend he had to resort to taking in pupils. But he was only four years younger than Newton. Could not Newton think of anything worse to say, or did Flamsteed's decorum prevent him from mentioning other 'ill names'?
4. *DNB* 1973, vol.5 p.842.
5. Flamsteed 1695. Quoted in Hoyt & Schatten 1997, p.24.
6. Quoted in Maunder 1922, p.143–44.
7. Parker 1975, p.50.
8. Halley 1716, p.423.
9. Wilson 1826, p.280.
10. Wilson 1774, p.14, 17.
11. Wilson 1783, p.154.
12. Swinden 1727, p.132.
13. Kant, I. 1981, p.165.
14. Kant, I. 1981, p.288.

Chapter 11

1. Asimov 1975, p.273.
2. *Edinburgh Review* 1803, vol.1, pt.2, p.431.
3. Brewster 1830, vol.6, p.618.
4. *Gentleman's Magazine* 1787, pp.636, 645.
5. Grant 1966, p.221. Still, that the interior of the Sun might be cold has had an enduring fascination. As late as the 1950s a wealthy German businessman, Gottfried Bühren, set up two prizes, each worth 25,000 DM. One prize was for disproving that the Sun was cold and covered with hot clouds, the other one for proving that the interior of the Sun is several million degrees hot. At that time the German astronomical society was short of funds and it was decided that three astronomers jointly prepare a paper for the first prize with the view that the money would go to the society. By the time the prize committee awarded the prize Bühren had a change of heart, did not pay up and took the matter to court. This case, like that of Dr Elliot's, had to be abandoned due to the plaintiff's death, this time from a heart attack following a car crash. Nevertheless, the prize was paid out to the society. It is now administered as a fund for travel grants to young astronomers.

Chapter 12

1. Wolf 1861, p.18. My translation.
2. Wolf 1861, p.18f. My translation.
3. Sabine 1852, p.175.
4. Sabine 1852, p.177.
5. We still talk of Wolf numbers, or Zurich sunspot numbers. The International Sunspot Number

is now computed every day in Brussels as a weighted average of several observations taken at different places on the globe adjusted to the observers' individual bias, their different instruments and the observing conditions. Americans compute their own sunspot numbers. The Boulder Number is substantially greater than the International Sunspot Number.

6. Visit *www.ngdc.noaa.gov* — home page of the National Geophysical Data Centre, or *www.science.nasa.gov/ssl/pad/solar* General information can be found on *www.sunspotcycle.com* and *www.ips.gov.au/papers* Another useful site is *www.spaceweather.com* which has excellent daily data of interplanetary conditions.

7. Campbell 1968.

Chapter 13

1. Kidwell 1984, p.537.
2. 'May 18 eclipse' 1901, p.322.

Chapter 14

1. King 1979, p.283.
2. Comte 1875, vol.1, p.115. It is worth mentioning that Martineau condensed Comte's six volume work into two volumes and her English version was so much better than the original that, retranslated into French, it was used to popularize positivist philosophy in France.
3. Quoted in Gillispie, 1972, vol.6, p.29.

4. Maunder 1894. The magazine *Knowledge* was similar to *New Scientist* and sported the motto 'Simply worded — exactly described.'
5. Clerke 1894, p.206.
6. Maunder 1922.
7. Eddy 1976. Some of Eddy's arguments *for* the minimum: telescopes were reasonably good by then; many good astronomers and amateurs were watching the sky; Far East naked eye observations of the period are also missing; records show that there were few auroras; several astronomers mention lack of spots (e.g. Cassini, Flamsteed, De La Hire); the corona was not conspicuous since eclipse descriptions often don't mention it.

Arguments *not* supporting the existence of the minimum: literature was thin in those days (absence of evidence is not evidence of absence); the publication of the *Rosa Ursina* may have discouraged sunspot watching, and the random nature of spots meant that they were not recorded seriously.

8. Eddy 1976, p.1200.

Chapter 15

1. Stetson 1947, p.127.
2. In the spring of 2000 NASA launched *IMAGE* (Imager for Magnetopause to Aurora Global Exploration) to obtain global images of the magnetosphere

and in July of the same year the European Space Agency launched *Cluster II* to investigate the physical connection between the Earth and the Sun. It consists of four spacecraft flying in formation as vertices of a tetrahedron (a pyramid with a triangular base). It has already produced spectacular pictures of solar emissions.

3. Clerke 1902, p.160.
4. A daily forecast issued by the University of Alaska can be accessed at *www.gi.alaska.edu /cgi-bin/predict.cgi* and you can request to be alerted by email of the possible appearance of an aurora through the website *www.aurorawatch.york.ac.uk* Another helpful site is *www.gi.alaska.edu.php3*

Chapter 16

1. When such a substitute is used it is called a 'proxy' measurement. For instance, if you want to know whether you have gained or lost weight, but have no scales to weigh yourself, the trousers fitting tightly or loosely will serve as a good enough proxy. Substitutes or proxies are often used when direct measurements are not available.
2. Hoyt & Schatten 1997, p.164.
3. To the Maunder Minimum in the seventeenth century and the Spoerer Minimum around 1400, a Dalton Minimum has also been added for the years around 1800.
4. Eddy 1976, p.1199.
5. Hoyt & Schatten 1997, p.197.
6. Maunder 1894, p.175.
7. Lean & Rind 1998, p.3082.

References

Arago, D.F.J. 1842. Quels ont été les premiers observateurs des tâches solaires? In *Annuaire pour l'an 1842 présenté au Roi par le Bureau des Longitudes.* Paris.

—. 1855–58. *Popular Astronomy.* London: Longmans.

Aristotle. 1984. *On the heavens.* In J. Barnes (ed.) *The complete works of Aristotle.* Princeton University Press.

Asimov, I. 1975. *Asimov's biographical encyclopedia of science and technology* (new rev. ed.) London: Pan.

Bartha, L. 2000. *Magyarországi csillagászok életrajzi lexikonja.*

Braunmühl, A. von. 1891. *Christoph Scheiner.* Bamberg: Buchner.

Brewster, D. (ed.) 1830. *Edinburgh Encyclopaedia.*

Burton, R. 1621. *The anatomy of melancholy.* Oxford: H. Cripps.

Campbell, W.H. 1968. Correlation of sunspot numbers with the quantity of S. Chapman's publications. *Transactions of the American Geophysical Union.* 49:609f.

Clerke, A. 1894. A prolonged sunspot minimum. *Knowledge.* 17:206f.

—. 1902. *Popular history of astronomy during the nineteenth century.* (4th ed.) Edinburgh: A.& C. Black.

Comte, A. 1875. *The positive philosophy of Auguste Comte.* Trans. Harriet Martineau. (2nd ed.) London: Tübner.

Descartes, R. 1983. *Principles of philosophy.* Trans. V.R. Miller and R.P. Miller. Dordrecht: D. Reidel.

DNB 1973. *Dictionary of national biography.* (repr.), ed. Stephen, Sir L. and Lee, Sir S., Oxford University Press.

Drake, S. 1978. *Galileo at work.* University of Chicago Press.

Eddy, J.A. 1976. The Maunder Minimum. *Science.* 192:1189–1202.

Edinburgh Review. 1803. [Review article] 1:2.

Einhard the Frank. 1970. *The life of Charlemagne.* London: Folio Soc.

Elizabeth I. 1584. *Letters Patent.* Quoted in 1984. Raleigh and Roanoke. *British Library exhibition notes.*

Fabricius, J. 1611. *De maculis in sole observatis ... narratio.* Wittebergae.

Flamsteed. 1695. Unpublished letter in the Cambridge University Archives.

Galileo, G. 1953. *Dialogue concerning the two chief world systems.* Trans. S. Drake. Berkeley: University of California Press.

—. 1957. *Letters on sunspots.* Trans. S. Drake. In *Discoveries and opinions of Galileo.* New York: Doubleday Anchor.

—. 1989. *Sidereus nuncius*. Trans. A. Van Helden. University of Chicago Press.

Gassendi, P. 1657. *The mirrour of true, nobility and gentility*. Trans. W. Rand. London: Moseley.

Gentleman's Magazine. 1787. 57:2:636, 645f.

Gillispie, C.C. (ed.) 1972. *Dictionary of scientific biography*. New York: Scribner's Sons.

Grant, R. 1966. *History of physical astronomy*. (repr. ed.) New York: Johnson.

Guiducci, M. 1960. *Discourse on the comets*. In *The controversy on the comets of 1618*. Trans. S. Drake and C.D. O'Malley. Philadelphia: University of Pennsylvania Press.

Halley, E. 1716. An account of the late surprizing appearance of the Lights. *Philosophical Transactions*. 347:406–28.

Harriot, T. 1588. *A briefe and true report of the new found land of Virginia*. London.

Hooke, R. 1674. A*nimadversions*. London: J. Martyn.

—. 1676. *A description of helioscopes*. London: J. Martyn.

Hoyt, D.V. and K.H. Schatten. 1995. Observations of sunspots by Flamsteed. *Solar Physics*. 160:381–83.

—. 1997. *The role of the Sun in climate change*. Oxford University Press.

Huggins, M.L. 1889. Warren De La Rue. *The Observatory*. 12:244–50.

Huygens, Christiaan. 1689. *Cosmotheoros* (2nd ed.) Glasgow: Foulis.

Kant, I. 1981. *Universal natural history and theory of the heavens*. Trans. S.L. Jaki. Edinburgh: Scottish Academic Press.

Kepler, J. 1953. *Epitome astronomiae Copernicanae*. In *Johannes Kepler gesammelte Werke*. Vol.7. ed. M. Caspar. München: Beck.

Kidwell, P.A. 1984. Women astronomers in Britain, 1780–1930. *Isis*. 75:534–46.

King, H.C. 1979. *The history of the telescope*. New York: Dover (repr.)

Lean, J. and D. Rind. 1998. Climate forcing by changing solar radiation. *Journal of Climate*. 11:3069–94.

Marius, Simon. 1916. *Mundus Jovialis*. trans. A.O. Pickard. *The Observatory*. 39:367–81, 403–12, 443–52, 498–503.

Maunder, E.W. 1894. A prolonged sunspot minimum. *Knowledge*. 17:173–76.

—. 1900. *The Royal Observatory Greenwich*. London: Religious Tract Soc.

—. 1922. The prolonged sunspot minimum, 1645–1715. *J. Br. Astr. Ass*. 32:140–45.

May 18 eclipse. 1901. *The Observatory*. 24:321f.

Newton, Sir I. 1961. Newton to Locke, 30 June, 1691. In *The Correspondence of Isaac Newton*. Cambridge University Press.

THE ENIGMA OF SUNSPOTS

Parker, E.N. 1975. The Sun. *Scientific American.* 233 (9):43–50.

—. 2000 June. The physics of the Sun and the gateway to the stars. *Physics Today.* 53:26–31.

Sabine, E. 1852. On periodical laws discoverable in the mean effects of the larger magnetic disturbances. *Abstracts of the papers communicated to the Royal Society of London.* 6:174–78.

Sayce, A.M. (n.d.) *Babylonian literature.* London: Bagster.

Shirley, J. 1983. *Thomas Harriot.* Oxford: Clarendon.

Stetson, H.T. 1947. *Sunspots in action.* New York: Ronald.

Swinden, T. 1727. *An enquiry into the nature and place of hell.* (2nd ed.) London.

Theophrastus. 1894. *On winds and weather signs.* Trans. J.G. Wood. London: Stanford.

Virgil. 1986. Georgics. In *Virgil.* trans. R. Fairclough. Cambridge, Mass.: Harvard University Press; London: Heinemann.

Welsh, J. 1904. The second voyage to Benin. In Hakluyt, R. *The principal navigations, voyages, traffics, discoveries of the English Nation* (repr. ed.) Glasgow: MacLehose.

Wilson, A. 1774. Observations on the solar spots. *Phil. Trans.* 64:1–30.

—. 1783. An answer to the objections stated by M. De la Lande. *Phil. Trans.* 73:144–68.

Wilson, P. 1826. Biographical account of Alexander Wilson, M.D. *Trans. of the Royal Soc. of Edinburgh.* 10:279–97.

Wittman, A.D. and Z.T. Xu. 1987. A catalogue of sunspot observations from 165 BC to AD 1684. *Astronomy and Astrophysics Suppl. Ser.* 70:83–94.

Wolf, R. 1861. *Die Sonne und ihre Flecken.* Zürich: Orell Füssli.

Yau, K.K.C. and F.R. Stephenson. 1988. A revised catalogue of Far Eastern observations of sunspots (165 BC to AD 1918). *Quarterly Journal of the Royal Astronomical Soc.* 29:175–97.

Further Reading

Bray, R. J. and R. E. Loughhead. 1964. *Sunspots*. London, Chapman & Hall.
Burch, J. L. 2001 April. The fury of space storms. *Scientific American.* 284:86–94.
Chapman, A. 1998. *The Victorian amateur astronomer*. Chichester: Wiley.
Ellison, M. A. 1968. *The Sun and its influence*. London: Routledge.
Giovanelli, R. 1984. *Secrets of the Sun*. Cambridge University Press.
Gribbin, J. 1991. *Blinded by the light: the secret life of the Sun*. London: Bantam.
Kippenhahn, R. 1994. *Discovering the secrets of the Sun*. Chichester: Wiley.
Lang, K.R. 1995. *Sun, Earth and sky*. Berlin: Springer.
—. 2001. *The Cambridge encyclopedia of the Sun*. Cambridge University Press.
Nicolson, I. 1982. *The Sun*. London: Mitchell Beazley.
Odenwald, S. 2000 March. Solar storms: the silent menace. *Sky and telescope.* 99:50–56.
Phillips, K.J.H. 1992. *Guide to the Sun*. Cambridge University Press.
Stetson, H. T. 1937. *Sunspots and their effects*. New York: McGraw-Hill.
—. 1947. *Sunspots in action*. New York: Ronald.
Taylor, P. 1991. *Observing the Sun*. Cambridge University Press.

The first publications

Fabricius, J. 1611. *De maculis in sole observatis, et apparente earum cum sole conversione, narratio.* Witebergae.
Scheiner, C. 1612. *Tres epistolae de maculis solaribus.* Augustae Vindelicorum.
Scheiner, C. 1612. *De maculis solarib. et stella circa Iovem errantibus accuratior disquisitio.* Augustae Vindelicorum.
Galileo, G. 1613. *Istoria e dimostrazioni intorno alle macchie solari e loro accidenti.* Roma.
Scheiner, C. 1630. *Rosa ursina, sive, sol ex admirando facularum et macularum suarum phaenomeno.* Bracciani.

Index

Entries in bold indicate photographs

Photo Credits

With thanks to the following for photos on the pages indicated:

British Museum: 21
Judit Brody: 19, 107, 134
By permission of the President and Fellows of Corpus Christi College,
    Oxford. 113 (top)
NASA website: 113 (bottom), 124 (top)
The Royal Astronomical Society: 13, 31, 72 (bottom), 76, 82, 91, 102, 128, 129,
    136, 137, 145
Christoph Scheiner, *Tres Epistlae de Maculis Solaribuis Scriptae ad Marcum
    Welserum* (1612) 56 and *Rosa Ursina* (1630) 59
Science and Society Picture Library/Science Museum: 11, 34, 36, 40, 41, 44, 56,
    57, 58, 61, 67, 68, 74, 75 (top), 77, 78, 80, 94, 96, 122, 124 (bottom), 125, 127,
    130 (bottom), 132 (bottom), 133, 135, 142
U.S. National Solar Observatory/Sacramento Peak: 117, 124 (bottom), 125, 145
Wellcome Trust: 71, 72 (top), 85
Still Pictures: 120
Science Photo Library 116, 121